Lecture Notes in Mobility

The book series Lecture Notes in Mobility (LNMOB) reports on innovative, peer-reviewed research and developments in intelligent, connected and sustainable transportation systems of the future. It covers technological advances, research, developments and applications, as well as business models, management systems and policy implementation relating to: zero-emission, electric and energy-efficient vehicles; alternative and optimized powertrains; vehicle automation and cooperation; clean, user-centric and on-demand transport systems; shared mobility services and intermodal hubs; energy, data and communication infrastructure for transportation; and micromobility and soft urban modes, among other topics. The series gives a special emphasis to sustainable, seamless and inclusive transformation strategies and covers both traditional and any new transportation modes for passengers and goods. Cutting-edge findings from public research funding programs in Europe, America and Asia do represent an important source of content for this series. PhD thesis of exceptional value may also be considered for publication. Supervised by a scientific advisory board of world-leading scholars and professionals, the Lecture Notes in Mobility are intended to offer an authoritative and comprehensive source of information on the latest transportation technology and mobility trends to an audience of researchers, practitioners, policymakers, and advanced-level students, and a multidisciplinary platform fostering the exchange of ideas and collaboration between the different groups.

Gereon Meyer · Sven Beiker
Editors

Road Vehicle Automation 12

Springer

Editors
Gereon Meyer
VDI/VDE Innovation + Technik GmbH
Berlin, Berlin, Germany

Sven Beiker
Stanford University
Palo Alto, CA, USA

ISSN 2196-5544 ISSN 2196-5552 (electronic)
Lecture Notes in Mobility
ISBN 978-3-032-12373-2 ISBN 978-3-032-12374-9 (eBook)
https://doi.org/10.1007/978-3-032-12374-9

This Springer imprint is published by the registered company Springer Nature Switzerland AG
The registered company address is: Gewerbestrasse 11, 6330 Cham, Switzerland

Preface

More than 12 years after SAE first defined the six levels of driving automation for on-road motor vehicles in taxonomy J3016, introduced in the opening volume of the Road Vehicle Automation book series, this 12th edition once again marks an important milestone. Five of the six SAE levels have now moved beyond speculation: lower levels are firmly established, Level 3 functions are available for highway driving, and Level 4 shared passenger vehicles have become a familiar sight on public roads. Only the most ambitious approaches, targeting Level 5, remain immature. Discussions on "full automation" still revolve around what would be required for a robotaxi operating solely on cameras and AI to navigate safely in an unrestricted operational design domain (ODD). Whether the solution lies in stronger AI, greater public tolerance for failure, or a combination of both, the fact remains that Level 4 driving services based on multi-sensing, connectivity, and dynamic mapping continue to expand their ODDs daily in the USA and China, moving step by step toward the vision "anytime, anywhere, under all conditions" which may ultimately define Level 5.

The progress of road vehicle automation continues to raise important questions for academic researchers, industrial engineers, and policymakers, questions that this series and its companion conferences have consistently addressed. Some argue that the fundamental debates are largely settled, with attention shifting toward proprietary problem-solving and collaborative partnerships. Others contend that reaching Level 5, or indefinitely extending the ODD of Level 4, remains the greatest challenge, requiring significant technical progress beyond cameras and AI. At the same time, the conference itself is evolving, reflecting the reality that automated driving has matured from an emerging research area into an established industry, yet one that continues to face substantial technological hurdles.

The contents of Road Vehicle Automation 12 reflect this transformation. Marking the completion of the first dozen in the series, this volume compiles contributions from the Automated Road Transportation Symposium (ARTS), held in San Diego, California, in July 2024. These chapters are based on original content selected by the conference organizers or breakout session leaders and subsequently presented and discussed at the symposium.

What stands out is a shift in focus. Contributions increasingly address the practical deployment and operation of self-driving vehicles, whereas purely experimental or conceptual studies were far more common just five years ago. This year's chapters explore themes such as testing automated vehicles, training and engaging first responders, defining and applying ODDs, and philosophical as well as organizational questions concerning operations and business models. Collectively, they show how the conversation has matured toward understanding deployment at scale.

Reviewing and editing these contributions has been both a privilege and a learning experience for us as editors. We particularly value the breadth of disciplines represented, ranging from engineering and policy to social sciences and ethics. The perspectives presented here provide a rich foundation for ongoing discussion, including the unresolved question of how to meaningfully compare driverless vehicles with those driven by humans, a topic that continues to generate interest and debate. In this regard, we are especially grateful to Steve Shladover, Jane Lappin, and Valerie Shuman for their leadership and commitment to organizing the Automated Road Transportation Symposium. Their efforts sustain a vital forum for dialogue across research, industry, and policy, benefiting not only this book but the entire automated driving community.

We would also like to extend our sincere thanks to the team at Springer Nature for their editorial expertise and guidance, which has once again resulted in an outstanding publication in the framework of the Lecture Notes in Mobility series.

We hope you will find the insights in this 12th volume of Road Vehicle Automation as engaging and thought-provoking as we did. We also look forward to meeting all of you again at the new Automated Transportation Symposium in Tempe, Arizona, in November 2025.

September 2025

Gereon Meyer
Sven Beiker

Contents

Part IV: Vehicle Technology Development and Testing

Introduction: The Automated Road Transportation Symposium 2024

Steven E. Shladover[1](✉), Jane Lappin[2], and Valerie Shuman[3]

[1] University of California PATH Program, 1357 South 46th Street, Building 452, Richmond, CA 94804, USA
steve@path.berkeley.edu
[2] Blue Door Strategy and Research, LLC, Belmont, MA, USA
jane@bluedoorsr.com
[3] Shuman Consulting Group, LLC, Skokie, IL, USA
vs@shumangroupllc.com

Abstract. In 2024, the Automated Road Transportation Symposium was held in San Diego, California, its first time at this location. The plenary presentations and breakout discussions continued to provide the meeting participants with the most up-to-date and authoritative information about the current international state of development and deployment of road vehicle automation systems, retaining its standing as the essential global meeting for industry, government and research practitioners in the field. Based on the experience of a full year of driverless operations by substantial numbers of vehicles in several cities and of a high-profile incident that shook public confidence and created a crisis for one of the industry leaders, the participants had plenty to consider in the well-packed meeting agenda.

Keywords: road vehicle automation · road transport automation · automated vehicles · autonomous vehicles · self-driving vehicles

1 Overview

The 2024 Automated Road Transportation Symposium (ARTS24) was organized and produced by a large team of professional volunteers working under the auspices of the National Academies of Science, Engineering and Medicine (NASEM) Transportation Research Board (TRB). The meeting was organized to serve the participants' interests in understanding the impacts, benefits, challenges and risks associated with increasingly automated road vehicles and the environments in which they operate. It brought together key government, industry and academic experts from around the world with the goal of identifying opportunities and challenges and advancing Automated Driving System (ADS) research across a range of disciplines.

The symposium was held at the Hilton San Diego Bayfront from 29 July to 1 August 2024. The plenary sessions were scheduled for the full mornings of the second and fourth days and half of the morning on the third day. The afternoons of the first three days were devoted to full-length breakout sessions, and half-length breakout sessions

G. Meyer and S. Beiker (Eds.): ARTSymposium 2024, LNMOB, pp. 1–12, 2026.
https://doi.org/10.1007/978-3-032-12374-9_1

were held on the morning of the third day. Six parallel breakout sessions were held in each of the breakout time slots, for a total of twenty-four breakout sessions. In addition, ancillary sessions organized by affiliated organizations (the American Association of Motor Vehicle Administrators, Automated Vehicles Industry Association, FHWA, IEEE ITS Society, ITS America and the SAE On-Road Automated Driving Committee) and the TRB International Working Group on Automated Driving were clustered around the main program.

The breakout sessions were organized by committees of volunteers to address a wide range of topics. These were clustered into three thematic tracks to make it easier for attendees to identify the sessions of strongest interest to them:

- Policy
- Operations
- Safety.

The plenary and breakout session programs were planned and produced by the ARTS24 Planning Committee, which included a mixture of TRB volunteers and support staff from Noblis:

John Craig, Noblis; Raymond Gerte, Noblis; Jane Lappin, Blue Door Consulting (TRB Vehicle-Highway Automation Committee Chair); Hyungjun Park, U.S. DOT Intelligent Transportation Systems Joint Program Office; Steven Shladover, University of California PATH Program (and former chair of the TRB Vehicle-Highway Automation Committee); Valerie Shuman, Shuman Consulting Group, LLC and Chair, TRB CORVA Subcommittee; Egan Smith, U.S. DOT Intelligent Transportation Systems Joint Program Office; Edward Straub, SAE International; and Marisa Walker, Arizona Commerce Authority.

2 Keynote Talks

The plenary program began with welcoming remarks by Jane Lappin and Edward Straub, who set the scene for a thematic sequence of plenary presentations and panel discussions on the dominant theme of safety of automated driving. This was followed by a memorial tribute to Barbara Wendling, a leader in the automated driving industry who passed away earlier in the year.

Steve Kenner, the Chief Safety Officer of Cruise, gave the opening keynote talk to kick off the safety thematic sequence. He discussed the lessons about safety that Cruise had learned from the crash in which their vehicle severely injured a pedestrian, and explained how Cruise was revising their corporate culture and technical approach to try to ensure that such an event would not be repeated. He described their new cross-functional regulatory engagement team as part of their organizational culture change to emphasize working with rather than against the cities where they are testing and deploying their systems in order to build community trust through transparency. They created an internal safety oversight board and began engagement with others in the industry by joining the Open Mobility Foundation and AV Safety INFOShare, a new nine-company industry collaboration to share safety information being coordinated by MITRE and PAVE. He concluded by charging the audience members to collaborate on setting safety policies, to push the AV companies to emphasize safety and to work together to improve public attitudes and perceptions about AV safety.

Tom Tang, the Chief People Officer of May Mobility, stepped in for Dr. Edwin Olsen, CEO, when Dr. Olsen was called away to a meeting in Japan. Tom gave the keynote talk on the second day of plenary sessions, discussing how May Mobility is emphasizing public transit applications of automated driving under a limited range of conditions. Tom drew from his personal experience as a race car driver to animate his description of May Mobility's approach to scaling the company, managing teams, and safety assurance.

Brian Cronin, the Director of the Intelligent Transportation Systems Joint Program Office of the U.S. DOT, gave the final-day keynote talk, summarizing the wide range of activities that the U.S. DOT has been engaged in with regard to connected and automated driving, taking a deeper dive to describe the ITS JPO Automation Program, which addresses cross-modal research needs affecting vehicle automation systems at all automation levels in four focus areas:

- Assessing the feasibility of conducting a small-scale Cooperative Driving Automation pilot (focusing on Levels 0–2 with interoperable connectivity) and package the accomplishments for design and architecture.
- Identifying likely safety-relevant scenarios where human performance and interaction behaviors with vehicle automation technologies have a mismatch and develop corresponding human factors best practices.
- Cross-cutting research and resource development including design, planning, operations, and impact assessments informed by improved analysis, modeling, and simulation.
- Exploring recommended practices to address safe and efficient ADS integration into the transportation system.

3 Safety-Themed Plenary Presentations and Panel Sessions

ARTS24 took a new approach by organizing a substantial portion of its plenary sessions around the theme of automated driving safety, with a carefully structured sequence of content to show the importance of emphasizing safety considerations at each stage in the development and operation of ADS.

3.1 Panel Discussion on Safety Management Systems for Automated Driving: Learning from Commercial Aviation

Dr. Marjory Blumenthal from RAND moderated this discussion session engaging representatives from industry with significant experience with safety management systems (SMS) in commercial aviation: David Wichner from Waymo, Tanya Wenneman from Forterra and Joe Kolly from MITRE. David Wichner began with a general explanation of the history of SMS and the reasons it is important for the benefit of those audience members who were not familiar with the concept. The panelists then discussed the challenges of adapting SMS to the automated driving environment, in which the regulatory landscape is much more fragmented than it is in commercial aviation. The discussion also covered the importance of setting realistic risk targets that can evolve over time. This helps to make the risks more visible and comprehensible to decision makers, who can then make better-informed decisions about allocating resources to mitigate risks.

3.2 Automated Vehicle Safety Cases: Structure and Scope

Prof. Phil Koopman from Carnegie-Mellon University introduced the concept of safety cases, building on the prior discussion about SMS, since the safety case includes the SMS as well as substantial technical content that explains how hazards are identified and mitigated. The safety case builds the logical argument for safety based on the system design decisions that are taken to mitigate each risk as well as the simulations and experiments that are performed to verify that the risks have indeed been mitigated. Professor Koopman said that safety cases for automated vehicles must address at least four essential topics to support a claim of acceptable safety: (1) the particular claims of safety being made, including what acceptably safe means for the system being discussed; (2) a well-reasoned argument encompassing assumptions, defenses of the validity of the assumptions, and evidence that supports the claims; (3) reasons why a design team justifiably believes they meet the stated safety claims; and (4) reasons why external stakeholders should find the safety case credible. The structure of a sound safety case is likely to differ across design teams based on the specifics of the vehicle and its operational concept.

3.3 Driving the Future: Integrating SUNRISE's Safety Assurance Framework with SYNERGIES' Scenario Generation

Jordi Pont from Applus + IDIADA explained how two European Commission projects that he is leading, SUNRISE and SYNERGIES, are developing a safety assurance framework for European ADS developers. They are defining a comprehensive catalogue of scenarios to represent the range of hazardous situations that ADS are likely to encounter on European roads, and then developing a toolchain including simulations, proving ground tests and on-road tests to build a safety case. This would lead to the creation of a common analytical platform that ADS developers throughout Europe could use to make their respective safety cases.

3.4 How to Compare Automated Driving Safety with Today's Human Driving?

Jane Lappin moderated this panel discussion extending the safety theme beyond the design and development stages into determining how to use the data that are being collected from public road testing and operations of ADS to estimate the real-world safety that is actually being achieved. The panelists were selected to identify the data challenges that are commonly misrepresented when comparing human driving safety to ADS. The panelists were John Scanlon of Waymo, Kevin Kefauver of VTTI, Carol Flannagan of UMTRI and Eric Teoh of IIHS, who have all been active with ADS safety assessments and identifying how to create realistic "apples to apples" comparisons of the safety of ADS with the baseline safety of human drivers operating under comparable conditions. Accurate comparisons of ADS and human-driven vehicles require equivalence in crash definitions (crash rate numerator) and measurement of driving exposure (crash rate denominator). The panelists discussed their alternative approaches to the challenges associated with identifying the relevant data and processing it in ways that can produce realistic and unbiased safety comparisons.

3.5 What Data Should Public Agencies Collect to Assess Automated Driving Safety?

Steven Shladover moderated this panel discussion about how to determine the appropriate data for public agencies to collect from the developers of ADS in order to be able to obtain realistic assessments of the safety of their operations. The panelists were Miguel Acosta from the California DMV, Tillie Chang from the San Francisco County Transportation Agency, Mufaddal Ezzy from Aurora and Jeff Farrah from AVIA, providing a mix of state, local and industry perspectives. They explored the natural tensions that arise between public agencies interested in having access to more comprehensive data, while industry is concerned about the resources associated with providing data and with ensuring the protection of confidential business information. All agreed that it will be necessary to converge on agreement among the relevant stakeholders to define consistent standards for reporting the safety-relevant data in order to earn public confidence in the safety of AV operations.

3.6 Communication and Automation: Strategies for Engendering Safety

In this session, the safety sequence was concluded with a compendium of short presentations by Laura Fraade-Blanar from Waymo, Alexandra Mueller from IIHS, Pnina Gerson from MIT and Annika Larsson from Volvo Group about how best to communicate with the public about the safety of all levels of driving automation. For the lower levels of automation, they discussed the public misconceptions about what level of automation is actually available on production cars today and the related risks of drivers over-trusting the systems to perform correctly under a wide range of conditions. The discussion highlighted the need for clearer communication around the current availability of AVs and proper use of ADAS, including clear definitions, explaining the capabilities of the automation, and explaining its added value. For the higher levels of automation, there are challenges about how to communicate the complexities of safety cases in understandable terms and how to demonstrate the absence of unreasonable risk from automated driving.

4 Plenary Panel Sessions on Non-Safety Topics

Three panel discussion sessions were convened to discuss topics of strong current interest not directly related to safety of driving automation.

4.1 Cities' Experiences with Automated Driving Systems on Their Streets

Marisa Walker moderated this panel discussion to advance from the comparable session at ARTS23, reviewing the experiences of the cities that are hosting or planning to host automated driving testing on their streets. The panelists were Rachel Castignoli from Austin, Steven Osborn from the District of Columbia, Tobias Brzoskowski from Hamburg and Andrew Glass Hastings from the Open Mobility Foundation. The different locations have different levels of regulatory authority and different kinds of relationships with the companies that are developing and operating the ADS. For example, Hamburg has regulatory authority and has developed a plan to use highly automated vehicles, shared and taxis, to expand the reach of the existing transit system, provide better service to people with mobility disabilities, and to reduce dependence on privately owned vehicles. They are working cooperatively with ADS developers and the transit system operator to pilot the ADS vehicles in service. The city of Austin, in contrast, has no regulatory authority, and negotiated with ADS developers on data-sharing and public safety issues. Because it cannot mandate safety reporting, Austin relies on reports posted by citizen eyewitnesses to achieve a more complete measure of ADS vehicle impacts on safety and traffic. Washington, DC, is working with a consultant to develop guidelines for ADS operators, with the added complexity of interacting with multiple agencies responsible for the safety and security of the nation's capital. The panelists emphasized the importance of data sharing between the cities and the companies – the cities informing the companies about locations that they should avoid (emergencies, special events, challenging locations) while hoping to receive real-time local road condition information from the companies. Where companies are not reporting relevant incident information to the cities, the cities can fall back on direct inputs from citizens (online dashboard established by Austin for that purpose). Open data standards could facilitate the desired sharing of information.

4.2 Paving the Way for ADS in Trucking: Public Sector Engagement and Enforcement

Finch Fulton from K&L Gates moderated this discussion among Zeke Reyna from Texas DOT, Bart Teeter from the Texas Highway Patrol and Michelle Chaka from Torc Robotics. The discussion focused primarily on interactions between automated trucks and first responder and regulatory agencies and on how to define a framework for the safety cases that the companies share with the regulatory agencies.

4.3 Wrap-Up Perspectives from Leading Trade Journalists

Jane Lappin moderated this final session with panelists David Welch from Bloomberg and Richard Bishop from Bishop Consulting. They discussed some of the recent trends in the AV industry that were evident in the discussions during the symposium. These included the crisis of confidence within the automated ride-hailing (aka "robotaxi") segment of the industry, where there are significant doubts about the economic competitiveness with human-driven ride hailing, exemplified by the cancellation of the plans for the Cruise Origin vehicle. This also led to speculation that Zoox could be heading more toward package delivery applications for their Amazon parent. They noted the high level of activity in China, motivated by government long-term goals for world dominance in the automotive industry, but leading to suppression of reports about safety incidents involving ADS in China.

Public transit applications of ADS are receiving more attention outside the U.S. than inside the U.S., with emphasis on vehicles smaller than full-size buses to serve intermediate levels of demand that do not need large capacity vehicles. Automobile manufacturers are looking at how to apply the technology for simpler use cases (such as motorway driving), with the hope of earning ongoing subscription revenues from users of the systems in private personal vehicles. The long-haul trucking applications continue to show promise in development but have yet to reach the milestone of regular operation without onboard drivers. By contrast, progress appears to have stalled with the sidewalk robots for local urban delivery.

5 Individual Plenary Presentations

Individual presentations were distributed across the plenary program in between the panel discussions to avoid Powerpoint fatigue from too long a sequence of consecutive presentations. These covered several different clusters of topics:

5.1 Presentations About Driving Automation Below Level 4

- Alexandra Mueller, Insurance Institute for Highway Safety (IIHS) – IIHS Ratings Program on Safeguards to Keep Drivers Engaged Using Partial Automation Properly
- Yoichi Sugimoto, Honda – Honda's Pioneering Level 3 Automated Driving System for Personally Owned Vehicles

5.2 Presentations About Specific Automated Driving Developments in the U.S.

- Kim Williams, Metropolitan Transit Authority of Harris County – The State of Transit Bus Automation
- Richard Bishop, Bishop Consulting – Trucking Industry Update
- Ariel Wolf, Venable LLP – U.S. AV Legislative Update

5.3 International Programs Relevant to Automated Driving

- Ryota Shirato, ITS Japan – Introducing Recent AV Developments in Japan
- Claire Wren, UK Department for Transport – Building a Framework for Automated Vehicles in the UK
- Stephane Dreher, ERTICO – ITS Europe – Driving Towards Automation: The European Perspective

6 Breakout Sessions

ARTS breakouts gather key experts from around the globe for more in-depth consideration of specific topic areas. The goal of the breakout sessions is to collaboratively answer the questions: *What needs to be true to make the AV vision become a reality? How can our research help drive progress year on year?* The 2024 program included 24 sessions that covered a wide range of specialized topics from across the field to enable this discussion for the industry as a whole (see program list below).

The primary findings from each day's breakout discussions were reported back to the plenary the following day. The combined summaries provided in these Daily Roundups distill the latest insights from across the industry. The 2024 discussions demonstrated the ongoing maturation of the early-stage AV industry, including:

- *AV Deployment Checklists.* As the AV industry gains more experience with deployments, we are learning key lessons in a wide variety of areas. Many breakouts reviewed and discussed this progress, with consistent emphasis on key ideas, such as the need to properly integrate concerns such as disability-forward thinking into the planning stages of AV projects, the need to effectively address procurement and business model issues, and the need to proactively educate policy-makers, regulators, collaborators, and the public on AV capabilities, opportunities, requirements and risks.
- *Shared AV Resources.* Along the same lines, we are starting to see the evolution of collaborative industry tools such as AV Safety InfoShare and usAVjobs.com, which provide the ability for AV developers to pool safety data to accelerate collaborative solutions to common challenges; and the ability for researchers, policy makers and the public to better understand the economic impact of AVs and learn about available workforce training, respectively.
- *Public Safety.* The interaction with first responders was a major cross-cutting topic in the 2024 breakouts. Several groups emphasized the need for standards in this area to enable fleet scaling, from safety best practices to standardized interactions with all types of first responders that AVs may encounter, as well as measures of effectiveness that will help guide responder training. Breakouts discussing trucking noted that public safety and commercial vehicle providers share a strong interest in using AVs to improve safety and emphasized that joint public and private determination of minimum performance is needed.

ARTS breakouts also continued to reflect the many different aspects and impacts of vehicle automation, with discussions ranging from energy use and traffic flow effects to transit integration, curb management and AI. In addition, ARTS participants continued to focus on making progress around topics such as safety assurance, human factors, and the translation of the rules of the road for use by AVs, as such issues remain critical to the effective operationalization of AVs as users of the road network.

6.1 ARTS24 Breakout Sessions

6.1.1 Policy Sessions:

- Beyond Safety: Are Automated Vehicles the Next Big Growth Engine?
- A Tale of Seven Cities: Wargaming AV Deployment
- Help Wanted: AV Education and Industry Partnerships for Blue, White and New Collar Jobs
- Disability-Forward AV Thinking: Changing Perspectives to Achieve AVs' Potential
- Grants Aren't the Only Answer: Using Federal Funds to Scale Technology
- Impact of New AI Governance on U.S. Business and Policy
- Curbs: How Will Autonomous Pick-up/Drop-off Unlock Better Curb Allocation?
- Energy and Demand of Connected and Automated Mobility Systems: A Problem-Solving Caucus

6.1.2 Operations Sessions:

- Federal Automation Grant Programs and ADS Grant Update

- Building ADS Industry Standards and Putting Them to Work: A Workshop on the Role of Standards in Urban Driverless Operations
- Business and Governance Models Over the World: An Open Forum on Success Factors and Expectations
- Autonomous Vehicles in a Data Wonderland
- Unravelling the Route Map to Future Automated Transit Bus Service
- Automating the Future: CDA Implementation Plan for Safer, Smarter Mobility
- Using Digital Twins and Digital Infrastructure to Advance Safety and Mobility
- Large Models for Automated Vehicle Impacts Analysis and Evaluation
- Automated Freight and Trucking
- National Collaboration Models Around Roadway Digital Infrastructure for Automation Research, Innovation and Deployment

6.1.3 Safety Sessions:

- Safety Assurance of Automated Driving
- Driver State Monitoring in Owned Automation
- An ODD and a Scenario Walk Into a Bar… and Ask for Safety Driven Validation
- Grounding AV Safety Approaches in Aviation Best Practices
- Automated Vehicle Rule of the Road Compliance: Theory, Practice and Policy
- AV/First Responders

7 General Cross-Cutting Observations

As the field of road vehicle automation has advanced and the level of knowledge of the issues has grown over the past several years, the areas of emphasis within the Symposium have continued to evolve. While many of the topics and themes were consistent with those discussed in prior years, this year had a stronger focus on the real-world ADS deployment implications, based on growing real-world experience with ADS.

7.1 Focus on Practical Considerations

The practical considerations that attracted considerable attention included topics such as:

- ADS interactions with public safety personnel/first responders (indicating the need for standards and training)
- ADS compliance with traffic laws (or variations of traffic laws adapted to ADS)
- Legal frameworks governing ADS testing and operations (in UK and US states)
- Assessing real-world safety of ADS
- Curb space allocations for loading and unloading passengers or goods
- Remote human support for ADS
- Development and implementation of technical standards governing ADS
- Industry consolidation and more realistic runways for rollout of new ADS features.

7.2 Infrastructure Support for Automated Driving

Infrastructure support for ADS has had limited visibility in prior years, but was more prominent this year, with a particular emphasis on digital rather than physical infrastructure. Much of this appeared to be related to the U.S. DOT work on planning for Roadway Digital Infrastructure with much broader applicability than ADS, but there is also strong attention to "cloud" computing for ADS in China. These trends have introduced a new distinction between "direct" and "network" connectivity. Examples of early progress were discussed in provision of incident and travel condition data from cities to ADS operators, particularly to help in routing ADS away from incident sites where some of the early deployments have experienced ADS interference with first responders.

7.3 Lessons for the Industry from the Crisis at Cruise

The Cruise crisis was a prominent topic of discussion because of its high visibility and serious implications for the industry as a whole. Steve Kenner's opening keynote talk raised awareness among all the attendees about how Cruise is responding, but everybody was concerned about what if it had happened to them. The implications were particularly salient regarding effective interactions between industry and public agencies at all levels of government, but the Cruise incident also raised consciousness about the importance of integrating SMS into the cultures of companies developing and operating ADS and learning how to earn public trust to effectively manage adverse events when they occur.

The Cruise crisis also brought increased attention to the need to understand how other industries that encountered safety crises were able to respond effectively in improving actual safety and the public perception of safety. Commercial aviation provided an excellent example, with its development of SMS and a mechanism for industry-wide sharing of safety lessons learned so that the entire industry can benefit from the lessons learned by each of its companies individually. The announcement of the AV Safety InfoShare initiative to enable the AV companies to share their safety-relevant experience was one of the salient take-aways from ARTS24.

7.4 Relationship of ADS to Driving Assistance Systems

ARTS24 devoted more attention to topics associated with the lower levels of driving automation than previous symposia in this series, as industry participants who found it too difficult to reach Level 4 ADS have shifted their focus to how they can use ADS technologies to enhance driving assistance. At the same time, there was also discussion about how the knowledge gained from the growing proliferation of driving assistance systems in topics like perception, human factors, driver monitoring and functional safety can be applied to improving ADS performance and safety.

7.5 Public – Private Collaboration

The increasingly evident challenges to developing and deploying ADS in real-world operations led both public and private sector organizations to a growing recognition of the need to work together rather than at cross purposes to achieve shared goals of improving road traffic safety and efficiency. This recognition was more evident in the discussions than in previous years, since progress toward deployment and increasing public interest in using and sharing their road space with ADS will depend on building mutual trust and confidence.

7.6 Challenges in Comparing AV and Human Driving Safety

The growing fleets of AVs testing and providing transportation services on public roads have generated real-world data that can start to shed light on the question of how ADS will affect traffic safety. The initial studies using those data generated serious discussions about how the ADS data should be compared with the human driving baseline. This has led to a broad recognition across the stakeholder communities that more serious attention needs to be devoted to defining consistent methods for collecting and analyzing the data to characterize ADS and human driving safety under comparable conditions to provide valid comparisons.

7.7 Economic Challenges

The recent consolidations within the AV industry have highlighted the economic challenges that it faces in trying to develop and deploy systems that have a viable business case. The initial focus on shared-ride passenger services in urban ride-hailing has been found too challenging by most who have tried to turn it into a profitable business because of the combination of high costs (for both development and operation) and limited revenue potential (because of competition from under-paid human ride-hail drivers and passenger reluctance to share their vehicle space). At the same time, the automotive OEMs have also been trying to determine how best to apply their limited resources on automation technology to enhance the capabilities of the driving assistance features in the much larger volume of private personal vehicles that they sell. The decision by GM to terminate the Cruise Origin vehicle development brought that into focus at ARTS24, and of course their subsequent decision to abandon the ride-hailing market has only reinforced the significance of the challenge. This has led to more interest among the remaining industry participants in focusing initially on narrower niche applications that are less technically challenging and in smaller market segments such as fixed-route public transit and off-road applications in industrial, mining, agricultural and forestry industries. After they have refined the technology to work well for these applications and have generated revenue from customers they will be better equipped to enhance the technology to operate throughout the more complicated general urban road environment.

7.8 Workforce Concerns

The concerns about workforce impacts of ADS have grown to the point that they stimulated an industry response in the formation of the US AV Jobs Coalition as a collaboration among ADS industry leaders and regional partners to develop job training and retraining programs and to collect research findings on the likely employment impacts of ADS. This is bringing the workforce issues to higher national visibility and should be providing information that can be useful to support the development of national, state and local ADS policies.

7.9 Risk-Based Approaches to Regulation

The growing concerns about societal risks associated with artificial intelligence have motivated international interest in focusing on risk-based approaches to defining regulations for systems such as ADS that rely on artificial intelligence. This can be seen as an alternative to more traditional regulatory approaches originally developed for much simpler technologies but that are not well suited to dealing with the complexity of ADS.

Part I: Public Sector and Policy Activities

Safety Validation of Automated Driving Systems Driven by ODD and Scenarios

Xavier Delache(✉) and Elsa Lanaud

French Ministry of Transportation, 1 Place Carpeaux, 92055 La Défense Cédex, France
x.delache@gmail.com, elsa.lanaud@developpement-durable.gouv.fr

Abstract. Safety verification and validation of automated driving vehicles, and more generally automated road transport systems, are driven by various approaches to demonstrate that these vehicles and systems will be free of unreasonable safety risk for their occupants and other road users, within their operational design domain (ODD). Among the approaches, scenario-based methods are gaining increasing interest in research, market developments and regulation. The session mainly discussed how the definition of a given ODD on one hand, a more generic scenario approach on the other hand, could be efficiently combined to support claims that these systems are designed to respond safely to reasonably foreseeable situations within the ODD.

Keywords: automated driving system · verification and validation · ODD framework · scenario approach

1 Introduction

A special breakout session on safety for verification and validation purposes was held in ARTS 24 in San Diego, organized by Gil Amid and Sagar Behere (Fortellix) to discuss: *An ODD and a Scenario Walk Into a Bar... And ask for Safety Driven Validation.* Through presentations given by a panel of experts, various aspects of safety verification and validation processes and methods were shared during the session. A more in-depth discussion was also enabled with the audience on the link between ODD and scenarios, their rationale and complementary purposes.

Presentations at the session were made by: Gil Amid[2], Joel Bernier[3], Chris Bartholomew[4], Xavier Delache[1], Sid Ram[5], Siddartha Khastgir[6], Sagar Behere[2], Narayanan Kidambi[7], Jeff Wishart[8]. This paper is based on their presentation, though they are not engaged by the summary provided below.

1. French Ministry of Transportation, 2. Fortellix, 3. Nuro, 4. SAE, 5. Gatik, 6. University of Warwick, 7. National Highway Traffic Safety Administration, 8. Science Foundation Arizona

G. Meyer and S. Beiker (Eds.): ARTSymposium 2024, LNMOB, pp. 15–23, 2026.
https://doi.org/10.1007/978-3-032-12374-9_2

1.1 Context

The breakout session originally came up from the perspective of several initiatives on operational design domain (ODD) frameworks, which were debated under the United Nation Economic Commission for Europe (UNECE) Forum for Harmonization of Vehicle Regulations (WP.29), Working Party on Autonomous/ Automated and Connected Vehicles (GRVA) fora, the most relevant to date being the integration document developed by Functional Requirements for Automated Vehicles (FRAV) and Validation Methods for Automated Vehicles (VMAD) groups, which formulates recommendations and guidelines for future regulatory developments [1]. In particular, the integration document gathered methods from current state-of-the-art on safety demonstration and evaluation of automated road vehicles and systems. This includes numerous works on safety assurance starting from ODD considerations even descriptions, and scenarios derived from ODD frameworks but there is a need to specify the role of both ODD and scenarios in demonstrating and assessing safety for automated driving systems.

The ODD framework seems to be a key point of the safety assurance framework but extremely tied to the concept of scenarios, which might need to be designed and modeled through a scenario-based approach [2]. The link between an ODD and a process through which scenarios are derived is drawn from the notion of representativeness, rather coverage when it comes to a pragmatic way of assessing the match between an ODD defined by the manufacturer and the operational driving environment in which the system is going to be deployed. Scenarios are also intended to establish a link between methodological approaches and testing procedures, which means to define metrics to assess system's behaviors and finally evaluate it with regard to pass-fail criteria [3].

1.2 State-Of-The-Art on the Scenario-Based Approach

A starting point to an automated driving system (ADS) safety validation is based on the ability to demonstrate that the vehicle in which the ADS is integrated meets safety requirements and is free of unreasonable safety risk for its occupants and other road users. This prerequisite, before considering test methods, which are developed under GRVA New Assessment Test Methods (NATM) papers [1], will lead to think about methodological aspects to generate and feed scenarios applicable to a specific ADS defined on a unique ODD.

In literature, the scenario-based approach plays an important role in safety demonstration. The European regulation from 2022 related to the type-approval of fully automated vehicles and their automated driving systems gives a central place to scenarios for safety demonstration purposes [4]. The European Joint Research Center's interpretation document published in 2024 elaborates on the scenario generation process as part of safety demonstration activities [5]. This approach stands as the starting point of verification and validation activities, which imply turning to testing procedures by applying an ODD framework to this methodological scenario approach [5].

Back to conventional design-validation activities and the V-cycle model, which requires to cover all development and validation phases of a system, safety demonstration is considered as a process combining steps for specifying functions and their safety requirements; and risk analysis stages [6]. When it comes to risk analysis, two different approaches can be combined:

- a deductive approach, which consists of identifying dangerous situations at the global system level, assessing the risk for the overall system based on its performance results and passing on from the overall system to subsystems and components;
- an inductive approach, consisting of analyzing the failure modes of components and sub-systems, and analyzing the results in terms of induced dangerous situations and performance inadequacies to the overall system.

That said, the search for coverage and representativeness is a major challenge to ensure the greatest (see sufficient) coverage of risk analysis. Scenarios play an important role in responding to this search [7]. The main expected contribution of the scenario approach is to avoid the omission of certain types of scenarios in quantified approaches resulting from traditional risk analyses. The robustness of the overall approach (scenario generation / quantified risk analyses) should therefore be based on the principle according to which scenarios resulting from the methodological approach are systematically included in the quantified analyses, even if it means qualifying them as "implausible" or to describe their consequences as "not serious", aiming at reaching the most reduced space of "unknown scenarios".

The scenario-based approach is a first major step for safety validation but doesn't stand alone: the ODD in which a specific ADS will be tested (or deployed) is a cornerstone. When considering the ODD, depending on the specific system under evaluation, there is a parallel with two notions: the operational design domain of the ADS itself and the operational domain on which the ADS will be deployed. The first one is directly related to the manufacturer's design whereas the second one is more related to the combination of a defined ODD with a specific route or zone with its specificities. The ODD framework [8] as mentioned in the literature therefore make reference to this very complex environment, including other related concepts such as behavioral competency and capabilities, which are linked to system's responses.

More generally speaking, safety demonstration activities, in particular verification and validation processes will consume scenarios, which will be an input to validation procedures [2]. In that perspective, building a complete structure for describing scenarios, consistent with existing literature on the ODD framework or object and event detection and response (OEDR) approach. This description structure aims at breaking down the space of possible scenarios into coherent axes, allowing to infer scenarios through the combination of these axes. Those axes are directly linked to the ODD description, which induce a pillar of transferability of attributes between ODD and scenarios [9].

The interdependency but also differences between the ODD and scenarios, be it at methodological or testing level, needs to be further explored.

1.3 Objective of the ARTS 24 Breakout Session

The aim of a specific breakout session was to organize a more in-depth discussion bringing experts' views on ODDs and scenarios, especially their close connection when it comes to system validation. Speakers were representative from public authorities, industry, and research institutes.

The objective of the session was to organize a discussion around an ongoing topic currently subject of international work and research. The subject of ADS safety validation is one of the main open items for large scale deployment as there are pilots all around the world and early deployment in a few areas in the United States of America and China.

The question at stake was mainly to better understand the roles ODDs and scenarios play for safety verification and validation [10], especially ODD specification, qualification and certification of an ADS as well as the meaning of ODD coverage and metrics associated. When it comes to scenarios, what are the links with the ODD, namely for testing purposes? How to systematically generate and derive a set of scenarios from a given ODD and how to qualify scenario coverage to that ODD? Another important aspect for safety verification and validation is the gap between technical specifications and operational implementation, including to consider behaviors and other environmental conditions both in ODD definition and scenario generation and selection.

At the end of the discussion, the intention was to better consider the relationship between ODD definition and the use of scenarios for systems' design and validation and, more precisely, to better consider how scenario generic approaches could be implemented to a specific system and its ODD.

2 Summary of the Discussion

The session was organized in two parts. Firstly, speakers gave their views on how to consider the ODD framework and scenarios for safety driven validation. Then, a round table discussion was held, moderated by speakers to discuss and challenge the positions, views and see how to handle the complementarity between ODDs and scenarios.

2.1 Rationale of ODD Frameworks: Taxonomies

The starting point of the discussion was to properly define terms and definitions around the ODD framework, especially the ODD itself. As an introduction, Gil Amid as the organizer reminded the definition of an "operational design domain" taken from SAE J3016 [10]: "operating conditions under which a given driving automation system or feature thereof is specifically designed to function, including, but not limited to, environmental, geographical, and time-of-day restrictions, and/or the requisite presence or absence of certain traffic or roadway characteristics". In other words, an ODD is a sort of n-dimension box in which the ADS is designed to safely operate, dimensions including visibility conditions, environmental conditions such as rain, lighting, breeze, infrastructure such as highway, roundabouts.

Among the presentations, one was dedicated to ODD standardization [11], which encompasses work from SAE, ISO, IEEE, BSI or UL (non-exhaustive list). Especially, J3259 from SAE makes the link between all existing literature on the ODD from taxonomy elements to attribute classification for a consistent description. In that perspective and given its previous definition, it reminds that an ODD is not a description of the driving environment or even a description of capabilities or performance guarantee. An ODD is a design element of the ADS defined by the manufacturer, characterized by attributes and elements, those harmonization among international pieces of work will lead to facilitate understanding and awareness among different stakeholders.

2.2 ODD Considerations for Use: Specifications

An ADS (feature to be precise) is designed to operate in one ODD by definition, which means one change in the ODD is a design change for the ADS feature. To date, there are numerous works on taxonomies but there is a need to move forward on how to use ODD in wider verification and validation processes, which leads to consider coverage and its issues brought.

Among the presentations, one attempted to show that beyond taxonomies, ODD is a design specification but ODD definition is not only ODD taxonomies or a list of attributes [12]. Moreover, considering that ODD and behaviors are coupled is a mistake as well as considering that everything is ODD too (internal state, scenarios). As an example, it is not because something is important that it will lie as an ODD attribute, meaning that there is more than just ODD definition for the purpose of safety verification and validation.

When it comes to specifications, we see easily that somewhere comes the "out of its specification", which means an "ODD exit" for the ADS [10]. The breakdown in the state of the ADS is also part of the verification and validation process, leading to consider other notions beyond the ODD itself, more on an assessor's point of view.

2.3 Verification and Validation of an ADS: Modeling ODD Coverage

The link between ODD and scenarios has been brought from a modeling approach to qualify and quantify the ODD defined by the manufacturer and ADS's behaviors. Among the presentations, some of them addressed the link with scenarios derived from the ODD [13, 14], metrics definition to assess scenario coverage [15], and the manufacturer's safety case [16], and research involvement to derive functional scenarios for testing purposes [17].

Firstly, [10] reminded the approach to which ODD is modeled: define a set of ODD specifications, then model the ODD and incorporate it into scenarios verification and validation flows, and gather metrics.

[13, 14] presented how scenarios are mapped from the ODD, especially how scenarios are derived from an ODD roadmap (road collector, standardized intersection and environmental conditions). Development will proceed per ODD element to provide deterministic requirements that can be verified and validated across system levels by organizing scenarios at functional, logical and concrete levels. That being said, the set of selected and parametrized scenarios will be subject to assessment in order to see whether it gives a good and sufficient coverage of reasonably foreseeable situations in the ODD.

The notion of completeness is contradictory insofar on one hand the aim is to demonstrate that all scenarios are taken into account and on the other hand the number of scenarios needs to be finite [15]. The good question is therefore to ask for sufficient coverage, which is tractable and measurable. The notion of sufficient coverage is intimately linked to the notion of risk, which can evolve as the system's maturity is increasing. Linking scenario coverage to risk makes a clear parallel with quantitative approaches such as the coverage map [15]. Quantified approaches aim at objectivizing ADS's safety, which would be by essence a comprehensive work.

In the same idea, the safety case approach came into place as an enabler to complement the grey zone between ODD definition and scenarios. As presented in [16] a safety case is defined as a structured argument, supported by a body of evidence, that provides a compelling, comprehensible, and valid case that a product is safe for a given application in a given environment. The safety case approach, closer to a concept, is based on the combination of processes, procedures and results to demonstrate and assess a system's safety. This constitutes a global evaluation of the developer's organization, the system's design and its performances. Among this structured argument lies the scenario-based testing approach, in which are detailed driving assessment metrics and evaluation methodologies, respectively aiming at sharing what is measured when testing and how to use the measurements to make an assessment of driving safety.

[17] presented the current status of NHTSA research on scenarios, especially for testing. It gave the starting point from a regulatory perspective under requirements: within its ODD an ADS must address crash-imminent scenarios, pre-crash scenarios and perform the entire dynamic driving task. When assessing a specific set of scenarios for a specific application, the question raised is how to address the infinite permutations of scenario features? The method developed will have to consider integrating the ODD into a scenario identification methodology. This approach lies on three layers: ODD definition done by the manufacturer, OEDR definition brought by the operational environment and maneuvers coming from system's capabilities. This approach reminds that ODD and scenarios are linked when evaluating and assessing an ADS and tries to address unique challenges attached to the fact that an ODD affects applicable scenarios, which are not applicable to every ADS use cases.

2.4 ODD Considerations to Derive Scenarios: Regulatory Approach

The driving scenario approach is a key part of automated systems' safety evaluation and validation all around the world. Regulatory aspects, left apart from previous presentations are crucial when it comes to linking the designer's approach and processes with requirements for assessment, be it for self-certification or type-approval applications.

As introduced, the scenario-based approach is part of the new European regulation related to the type-approval of fully automated vehicles' automated driving system, to cover reasonably foreseeable risks in the ADS's ODD [4]. Currently, a new international regulation is under development under UNECE fora, which is intended to be applicable for both self-certification and type-approval regimes [18].

The last presentation highlighted a regulatory approach, in which scenarios are the backbone on the safety demonstration process [19]. Since 2022, French doctrine has focused on scenario generation, based on an architecture of descriptive layers and on the combination of axes and parameters. The combinatory-based approach has the capacity to generate a wide diversity of scenarios, and to handle both operational design domains (ODD) and operational domains (OD). It can be applied to the vehicle capabilities, or to a broader system integrating remote capabilities or to more specific risks specific to dedicated zones or routes for example. This approach considers the scenario generation and production as an iterative process: the combinatory-based approach can be continuously fed by the identification of new descriptors or scenario parameters, and capitalize on feedback to continually self-update.

However, this scenario generation approach needs to be complemented with tools allowing its quality to be assessed by a third party. This involves, in particular, evaluating the capacity of a scenario approach to represent the diversity of scenarios and to (virtually) reach the limits of the reasonably foreseeable space. Secondly, there is a need to ensure that system's response is free from unreasonable risks. This third-party view could only reasonably be based on a limited number of scenarios, on which the system's response would at least be described, and if necessary, evaluated, by simulations, tests or evaluations on open roads [20].

To develop this third-party regard on scenario generation processes, two approaches appear possible: one, deductive, would extract from the manufacturer's generation process, the most representative or most severe scenarios. The other approach, more inductive, would develop, independently of the manufacturer's scenario generation process, lists of "essential" scenarios. The first approach has the advantage of being, by design, consistent with the ODD; the second requires matching "external" lists to the system, with its ODD, but it seems less intrusive in the designers' practices.

When choosing between these approaches, care must be taken to ensure that neither the assessor nor the regulator replaces the industry in the choice of the scenario space, directly derived from the manufacturer's ODD.

A hybrid approach can then be designed, in which the external assessor and the designer interact from a reduced number of families of functional scenarios established by the regulator, in order to extract the concrete scenarios most faithfully representative of the system's ODD and the reasonably predictable envelope defined by the system's designer [21]. This hybrid approach could use some randomization in the selection process from generic functional scenarios to localized concrete scenarios.

3 Conclusions

The breakout session was organized in two parts: presentations from experts and discussion with the audience moderated by speakers in short groups. The second part of the session proposed to exchange with the audience based on concepts and ideas put forward by presentations. The diversity of stakeholders covered various points of view, inter alia authorities, industry, academia.

First, the concept of ODD was presented and discussed, referring to related standards using various terminologies. Defining ODD needs to clarify the purpose of this definition.

When it comes to safety verification and validation purposes, ODD coverage is a key criterion, which requires evidence in the safety case. External scenario lists or libraries provide useful references to support sufficient coverage demonstration. However, assessing coverage shouldn't rely on processes whose sole objective it to produce more and more scenarios. Besides, coverage being related to ODD definition, lists or libraries need to be derived into relevant scenarios within the ODD.

When it comes to communication and public education, regulatory and public perspectives may differ. This creates a need to explain ADS features' capabilities to consumers and users, including other road users who will interact with it, in a different way than how they are managed by the expert community.

Discussions illustrated the needs for further work on scenarios, as a key component of safety demonstration approaches, and a fruitful basis for interactions between the manufacturer's safety case, the requirements from the authority, and an assessor, be it an authority's technical service or another external / internal assessor.

References

1. Informal document GRVA-18–50, Guidelines and recommendations for ADS safety requirements, assessments and test methods to inform regulatory development, 18th GRVA, Agenda item **4**(c) (2024)
2. French Ministry of Transportation, Safety demonstration of automated road transport systems (ARTS) – Expected contributions of the driving scenarios, Methodological document (2022)
3. Donà, R., Ciuffo, B.: Virtual testing of automated driving systems. A Survey on Validation Methods, in IEEE Access **10**, 24349–24367 (2022). https://doi.org/10.1109/ACCESS.2022.3153722
4. Commission Implementing Regulation (UE) 2022/1426 of 5 August 2022 laying down rules for the application of Regulation (EU) 2019/2144 of the European Parliament and of the Council as regards uniform procedures and technical specifications for the type-approval of the automated driving system (ADS) of fully automated vehicles
5. Joint Research Center, Interpretation of EU Regulation 2022/1426 on the Type Approval of Automated Driving Systems, Technical report (2024)
6. French Ministry of Transportation, Safety demonstration of automated road transport systems: articulation of the different activities around the "scenario" approach, Methodological report (2022)
7. Lanaud, E., Delache, X. : Regulatory Framework and Safety Demonstration Principles in France, In: Meyer, G., Beiker, S. (eds.) Road Vehicle Automation 10. ARTSymposium 2022. Lecture Notes in Mobility. Springer, Cham (2023)
8. Zhang, X., Khastgir, S., Jennings, P.: An ODD-Based Scalable Assurance Framework for Automated Driving Systems. SAE Technical Paper 2023–01–0574 2023). https://doi.org/10.4271/2023-01-0574
9. French Ministry of Transportation, Safety demonstration of automated road transport systems: contribution of driving scenarios – Scenario use for safety demonstration, Methodological document (2023)
10. Amid, G., Behere, S. : An ODD and a Scenario Walk Into a Bar… and ask for Safety Driven Validation. ARTS 24, Breakout session #201 (2024)
11. Bartholomew, C. : ODD Standardization Activity, ARTS 24, Breakout session #201 (2024)
12. Khastgir, S. : Defining ODD: Use, Misuse, Disuse, and Abuse, ARTS 24, Breakout session #201 (2024)
13. Ram, S. : Anatomies of ODDs, scenarios, and routes for middle mile autonomy, ARTS 24, Breakout session #201 (2024)
14. Bernier, J. : Strategies for mapping ODD elements to functional scenarios, ARTS 24, Breakout session #201 (2024)
15. Behere, S. : Scenarios and metrics for high ODD coverage, ARTS 24, Breakout session #201 (2024)
16. Wishart, J. : DDT Performance Assessment of an OEM AV, ARTS 24, Breakout session #201 (2024)
17. Kidambi, N. : Overview of NHTSA Research on ADS Driving Scenarios, ARTS 24, Breakout session #201 (2024)
18. WP.29, GRVA, Automated Driving System Informal Working Group (ADS IWG). https://wiki.unece.org/pages/viewpage.action?pageId=238223362

19. Delache, X. : A regulator enters the bar and asks for an innovative and sober cocktail of self-certification, national type-approval, and local scrutiny, ARTS 24, Breakout session #201 (2024)
20. French Ministry of Transportation, Safety demonstration of automated road transport systems: contribution of driving scenarios – From scenario generation to scenario selection, Methodological document (2024)
21. French Ministry of Transportation, Safety demonstration of automated road transport systems: contribution of driving scenarios – What contribution do lists of functional scenarios make?, Methodological document (2024)

Automated Vehicles and First Responders

Jason JonMichael[1](✉), Ximon Zhu[2], Jeffrey Wishart[3], Alex Rudin[4], Virginia Lingham[5], Nanette Schieke[6], and Ian Berg[7]

[1] Highly Automated Systems Safety Center of Excellence, U.S. Department of Transportation, Washington, DC, USA
Jason.JonMichael@dot.gov
[2] Numobility LLC, Atlanta, GA, USA
ximon@numobility.ai
[3] Arizona Commerce Authority, Phoenix, AZ, USA
jeffw@AZcommerce.com
[4] MITRE Corporation, McLean, VA, USA
arudin@mitre.org
[5] WSP USA Inc., San Diego, CA, USA
Virginia.Lingham@wsp.com
[6] Texas A&M Transportation Institute, College Station, TX, USA
n-schieke@tti.tamu.edu
[7] U.S. Department of Transportation, Cambridge, MA, USA
Ian.Berg@dot.gov

Abstract. The advent of automated vehicles (AVs) on public roadways is transforming the transportation landscape, presenting challenges for first responders. While AV deployments are currently limited to only a handful of cities, these regions are leading the way in addressing the complexities of AV interactions through collaboration between emergency services, local governments, and AV developers. As AV deployment increases, first responders are currently tasked with adapting their procedures and training to ensure public safety, passenger care, and effective incident resolution. This chapter explores the key areas discussed where first responders interact with AVs, highlighting the challenges, progress, and areas requiring further research and development.

Keywords: automated driving systems · first responders · law enforcement · interaction plans

1 Introduction

This chapter first provides a summary of relevant discussion from previous symposia and the literature. Then, the session panels and breakout discussions are synthesized. Finally, it discusses next steps and areas for further research.

G. Meyer and S. Beiker (Eds.): ARTSymposium 2024, LNMOB, pp. 24–37, 2026.
https://doi.org/10.1007/978-3-032-12374-9_3

1.1 Key Themes from Previous Symposia

As automated driving technologies have advanced – quickly in the eyes of some, and too slowly for others – their integration into smart communities and connected infrastructure systems became a central focus of industry, public agencies, and emergency responders. Previous editions of the Automated Vehicle Symposium (AVS) and the Automated Road Transportation Symposium (ARTS) have included discussions on this topic in their programs to share relevant research results, lessons learned from pilot deployments, and policy/standards developments.

From theoretical explorations in 2019 to practical feedback from real-world deployments in 2022, the discussions have evolved to address the multifaceted challenges and opportunities presented by AVs. Across these years, recurring themes such as enhancing emergency response capabilities, fostering multi-stakeholder collaboration, and advancing data-sharing frameworks have driven the development of actionable strategies to ensure safety, efficiency, and innovation in AV adoption. The following summaries highlight the key insights, outcomes, and research priorities from sessions held in 2019, 2020, and 2022.

1.1.1 Overall Trends 2019–2022

Across the years, sessions highlighted the importance of integrating AVs into connected infrastructure with a focus on safety, efficiency, and collaboration. Key recurring themes included:

- The role of AVs in emergency response and management.
- Collaborative frameworks for multi-jurisdictional and public-private partnerships.
- Need for adaptive research and policy frameworks to address emerging challenges.

The meetings emphasized automated driving technology's transformative potential while stressing the importance of stakeholder involvement to drive innovation and safety. By 2022, these efforts reflected a comprehensive understanding of AV integration, setting the stage for more seamless adoption in smart communities and enhanced emergency response systems.

Such trends over time have included:

- **From Theoretical to Practical**: Discussions evolved from conceptualizing AV use cases to addressing practical challenges and experiences.
- **Safety and Efficiency as Core Goals**: The potential of automated driving systems (ADS) to enhance safety and emergency response efficiency remains a consistent priority.
- **Growing Complexity**: As AV deployment expanded, the interplay between technology, policy, and practice grew more intricate, necessitating ongoing collaboration and adaptation.
- **Urgency of Actionable Solutions:** The transition from research to implementation underscored the need for concrete actions, including standardized protocols, training programs, and advanced technologies to support AV integration.

Across 2019, 2020, and 2022, recurring themes and evolving priorities emerged, illustrating the growing complexity and importance of AV integration into connected infrastructure systems, particularly in emergency response and public safety. Discussion of the evolution of some particular themes follows.

1.1.2 Emergency Response Integration

In 2019, attendees initiated discussions on the potential of AVs to enhance emergency management, emphasizing practical scenarios such as emergency dispatch, responder interaction, and public safety communications. Participants identified emergency responders as a "missing mode" in planning for AVs.

The 2020 symposium deepened the focus on safe interaction protocols between AVs and first responders. Key discussions revolved around AV behavior during incidents, traffic incident management, and the need for real-time data sharing between AVs and emergency services.

In 2022, first responders provided real-world feedback based on hands-on AV deployment experience. They identified practical challenges, such as accessing AV-generated data during emergencies, and emphasized refining AV-first responder protocols based on lessons learned.

Over the years, the discussions on this theme matured from theoretical frameworks to addressing real-world deployment challenges, emphasizing that AV technology must be actively shaped to meet first responder needs.

1.1.3 Data Sharing and Technology Evolution

In 2019, the session highlighted the need for robust communication technologies and securing infrastructure to enable safe AV operations and facilitate emergency response.

In 2020, participants explored data sharing opportunities, particularly during crash investigations, traffic enforcement, and incident management. Topics addressed included the need for reliable "black box" data access and collaboration frameworks to bridge gaps between ADS developers and emergency services.

The 2022 symposium stressed the importance of streamlined data exchange mechanisms, including standardizing data collection and sharing protocols. New technology advances, such as real-time data dashboards and enhanced AV sensors, were seen as opportunities to improve response efficiency.

Progression in discussions reflects increasing recognition of data as a critical enabler for AV functionality and emergency management. Effective data sharing requires industry collaboration, technological advancements, and policy innovation.

1.1.4 Collaboration and Multi-Jurisdictional Coordination

The 2019 session raised awareness of the need for interagency collaboration across transportation, public safety, and communication sectors to establish a unified AV integration framework.

In 2020, participants discussed the complexities of multi-jurisdictional coordination, highlighting organizational roles, stakeholder partnerships, and regional cooperation as critical factors in AV deployment.

In 2022, the session focused on collaborative mechanisms through public-private partnerships and inter-agency frameworks to bridge gaps in AV integration and incident management. It emphasized training for first responders and public safety officials to adapt to automated driving technologies.

As AV adoption expands, multi-stakeholder collaboration has become essential to navigating jurisdictional, operational, and technological challenges.

1.1.5 Policy and Research Priorities

The 2019 session identified the foundational research needs to enable AV integration into connected infrastructure, including security, communication technologies, and emergency response applications.

In 2020, participants began formalizing a research agenda, focusing on issues like crash investigation protocols, traffic enforcement adaptation, and AV-first responder interactions. Federal perspectives played a larger role in shaping regulatory frameworks.

Discussions in 2022 prioritized policy-driven research to guide technology deployment and stakeholder training. Practical feedback from deployments informed recommendations for future studies and policy adjustments.

Over time, there has been a clear shift from exploratory research needs to actionable policy and technology strategies, emphasizing evidence-based approaches to AV integration.

1.1.6 Expanding Stakeholder Engagement

In 2019, the focus was on engaging transportation and emergency management communities to address initial AV integration challenges.

The dialogue expanded in 2020, to include industry leaders, government officials, and researchers, fostering a broader, interdisciplinary perspective.

In 2022, the session included first responders, vehicle technology providers, academia, and public agencies in discussions, enriching the conversation with practical insights and diverse perspectives.

Broadening stakeholder participation has strengthened the alignment of ADS technologies with real-world needs, ensuring that deployment strategies are inclusive and well-informed.

1.2 Key Literature

In addition to the evolving conversations at previous symposia, research focused on ADS/first responder interactions has continued to grow over the past several years. While not a comprehensive review, this section identifies some of the foundational literature published over the last several years that has helped advance the field.

One thread of research has focused on the types of interactions that vehicles and emergency responders have with each other. The Crash Avoidance Metrics Partnership (CAMP) worked in 2018 to identify the most common scenarios through literature review and interviews with public safety officials [1]. Examples include responding to an incident; securing a scene; conducting traffic control, traffic stops, and checkpoints; handling abandoned or unattended vehicles; and passenger stabilization and extrication. These scenarios were then examined using hierarchical task analysis, leading to flowcharts detailing each step a responder performs in a given scenario. The report also discussed opportunities where ADS may have the potential to improve vehicle-responder interactions.

Another research focus has been on responder training needs. The Governors Highway Safety Association (GHSA) commissioned a 2021 study which developed curriculum recommendations and anticipated learning outcomes [2]. These topics included not only interacting with ADS-equipped vehicles, but also understanding the capabilities of various levels of driving automation, identifying technologies on the road, preparing for future deployment, and understanding and accessing data.

The CAMP and GHSA reports were both foundational to the recommendations developed by the Automated Vehicle Safety Consortium (AVSC), the *AVSC Best Practice for First Responder Interactions with Fleet-Managed Automated Driving System-Dedicated Vehicles (ADS-DVs)* [3]. Initially published in 2020, and revised in 2024, this document builds on the interaction use cases identified by CAMP and categorizes interactions as direct, indirect, or informational. It also defines a minimum set of elements that should be included in an ADS developer, manufacturer, or fleet operator's first responder interaction plans.

The aforementioned reports include multiple types of responders, including law enforcement, fire, and emergency medical services. A 2020 report sponsored by the National Institute of Justice includes some additional considerations specific to law enforcement [4]. For example, the report discusses cybersecurity risks, and use of ADS data for investigations.

2 Summary of the Discussion

The AV/First Responder session was an interactive, design-thinking session. For this discussion, it was imperative to have an inclusive representation of stakeholders, including infrastructure owner-operators, ADS developers and operators, State departments of transportation, American Association of Motor Vehicle Administrators (AAMVA), and first responders at both the city and state levels. Participating Federal stakeholders included the National Highway Safety Traffic Administration (NHTSA), National Transportation Safety Board (NTSB), and the Federal Law Enforcement Training Centers (FLETC).

The first responder community ultimately holds the keys to driving and generating change in AV interactions and data acquisition at the local level. First responders of course include law enforcement – those who patrol, observe, and enforce traffic safety on our roadways, and those who respond to the incident scene. Also included as first responders are firefighters, emergency medical services (EMS), traffic incident management, safety service patrols, and towing/recovery personnel.

2.1 Current State of the Practice for First Responder Interactions with AVs

Cities with active AV deployments are serving as test cases, offering critical lessons that could inform future nationwide integration of AV technology. First responders are benefiting from closer partnerships with the companies developing and operating these vehicles. Collaboration often includes workshops and training sessions where the AV developers work directly with police, fire, and emergency medical services to explain how their systems operate and how to interact with them safely during an incident. These sessions provide critical information on identifying whether the automated driving system (ADS) is engaged, using manual override features to disable the system, and understanding the external indicators that AVs use to communicate their status.

This direct engagement helps responders gain hands-on familiarity with AVs, a vital element to experiential learning required to proficiently respond during real-world incidents. First responders benefit significantly from field exercises where they can interact directly with AVs in controlled environments. The breakout session underscored examples of cities that have developed partnerships to conduct large-scale drills simulating AV crashes, battery fires, and stalled vehicles. These exercises not only improve technical familiarity but also help responders build confidence in their ability to safely manage AV-related incidents.

Additionally, these cities are taking proactive steps to simulate and rehearse potential AV-related incidents. Tabletop exercises and full-scale drills allow first responders to practice handling scenarios such as collisions, battery fires, and teleoperated AVs. These simulations often include representatives from AV companies, enabling immediate feedback and refinement of emergency protocols. Through these exercises, cities are not only improving the readiness of their first responders but also fostering a culture of collaboration that ensures continuous learning as the technology evolves.

Insights gained from the session emphasize the need for consistent communication between AV developers and emergency responders. Stakeholders have highlighted the importance of creating standardized emergency procedures across all AV manufacturers to ensure first responders know how to handle incidents regardless of the vehicle's design. For example, the session emphasized that while some AVs provide clear and visible external indicators of their ADS engagement status, others do not, leading to confusion during emergencies. Stakeholders agreed that a unified approach to AV signaling—similar to standardized and uniformed traffic signs and signals—could alleviate this challenge.

Local governments in these early deployment cities have played a pivotal role in facilitating coordination between AV developers and first responder agencies. By establishing task forces, advisory committees, or working groups, these cities provide platforms for stakeholders to share information, identify challenges, and develop solutions collaboratively. For example, local agencies work with developers to create emergency response protocols, ensure teleoperators are accessible during incidents, and address unique risks associated with AVs, such as electric battery fires or the lack of a human driver. These partnerships are laying the foundation for standardized practices that can eventually be adopted on a larger scale.

The session also drew attention to emerging issues like remote support and the need for more uniform and verifiable methods of data collection. Some AVs rely on remote assistants to provide support in emergencies, but delays in accessing these operators during incidents remain a critical barrier. Furthermore, first responders require timely access to vehicle data for incident investigation and reconstruction. Participants stressed the need for clear policies that balance public safety with data privacy concerns while ensuring first responders can retrieve essential information without unnecessary delays.

The lessons emerging from these cities highlight the value of proactive engagement, hands-on experience, and standardization. First responders in these regions are at the forefront of adapting to AV technology and the experiences shared by first responders underscores that they are not only adapting to the technology but also leading the way in shaping national best practices. Their experiences emphasize the importance of creating clear, consistent procedures and fostering strong partnerships that can be scaled nationwide. However, this collaboration remains localized, and much of the country has yet to benefit from the progress being made. Without broader dissemination of these lessons and more coordinated efforts across regions, there is a risk that preparedness will remain uneven as AVs expand into new markets.

While the regulatory landscape for AVs remains fragmented, these early deployment cities are also working to address the legal and procedural gaps that accompany the new technology. Questions about authority to disable AVs, liability in the event of system failures, and access to AV data during emergencies are being discussed and, in some cases, addressed through collaboration between stakeholders. These efforts are paving the way for a clearer regulatory framework that can support first responder needs while balancing the rapid pace of AV deployment.

Ultimately, the limited number of cities experiencing AVs are playing a vital role in preparing for a future where this technology is commonplace. Through collaboration with AV developers, local governments, and emergency services, they are building the knowledge, protocols, and training systems that will be critical as AVs deploy to new regions. By continuing to emphasize communication, shared learning, and proactive engagement, these cities are not only improving their own preparedness but also setting the stage for a more coordinated, national approach to AV-related incidents. The insights gained from these early collaborations will be invaluable for first responders everywhere, ensuring they are equipped to safely and effectively manage the challenges and opportunities presented by AVs.

2.2 Breakout Group Summary

The session proceeded with several breakout discussions on important topics: communications between AV operators and first responder organizations, education and training, and first responders interacting with AVs in traffic. Each discussion followed a "roses, buds, and thorns" framework focusing the conversation on existing good practices (roses), promising initial efforts that require further development (buds), and roadblocks to successful deployment (thorns).

2.2.1 Communications Between AV Operators and First Responder Organizations

In the breakout group focused on Communications between AV Operators and First Responders, several valuable new ideas were presented. The fact that some AVs are capable of recognizing human traffic controller hand signals was recognized as a positive and necessary step in bridging these communications. There was also recognition that law enforcement interaction plans (LEIPs), required by some States, provide an existing platform to convey needed information to first responders in advance of operations. However, it was noted that law enforcement are not the only responders to an incident or emergency with whom an AV operator may need to communicate. The AVSC Best Practice refers to their recommended document as a "First Responder Interaction Plan" rather than a LEIP.

Areas for improvement were identified as well. Many of these were suggested for the vehicle itself, but it was also mentioned that cities and first responders can have a role by broadcasting active response zones for vehicles to avoid. This balance shows that communication between AVs and first responders is a complex, multi-stakeholder problem that will require coordinated solutions. It was also noted that several cities have these systems implemented and AV operators are ingesting this data and applying geofences to avoid active response zones, but that this practice could be more widespread and improved upon in the future.

On the vehicle side, there were concerns over the ability of a first responder to move the vehicle and to know the appropriate mechanisms for doing so. This concern was especially prevalent for custom built vehicles where controls and processes may be less familiar to first responders. With this, it was expressed that first responders must have confidence in when the vehicle will or will not move. For example, if it began to move while a first responder was extricating a passenger, there would be significant safety risks.

Participants also noted that first responders must know how to get into a vehicle to provide care to a passenger, and that a passenger safety plan may help inform the AV operator, the first responder, and the passenger on the correct actions.

Generally, AV operators expressed that this information is provided in their LEIPs and first responder trainings, but there continue to be opportunities to streamline these communications and reduce the cognitive load on first responders during high-intensity events. This could include improving indirect communications so that a first responder does not need to approach the vehicle's window to interface with a remote operator or incorporating external human-machine interfaces for bidirectional communication.

Finally, this group discussed that, even once solutions to these communication issues are implemented, it will be critical to continue refinement and streamlining of these communications. Between the array of AV platforms in use, among other road users, avoiding oversaturation of information is key to ensuring that different emergency responders receive the necessary information, and that unnecessary information is filtered out.

2.2.2 Education and Training

The breakout group on "Education and Training for First Responders and AVs" brought together key stakeholders from AV operators, first responders, and state and local transportation agencies to discuss the current state of educational programs and training protocols designed to prepare first responders for interactions with AVs. Participants recognized the importance of comprehensive, up-to-date training materials and the need for ongoing collaboration between AV operators and first responder agencies.

Several case studies were presented, highlighting the proactive efforts of industry leaders and government agencies in developing educational resources and training programs. For example, PennDOT's AV Incident Response Plan (AVIRP) and its companion mobile application, the AV Incident Response Field Guide (AVIR-FG), were showcased as innovative tools that facilitate real-time communication and reporting during AV-related incidents. Similarly, Waymo's First Responder Program and Guide, which has been used to train over 15,000 first responders across the United States, was discussed as an example that aligns with industry best practices. These case studies are detailed further below.

While acknowledging the progress made thus far, workshop participants also identified areas for improvement and future collaboration. These included the need for incorporating AV training into existing programs such as traffic incident management (TIM) response training, and integrating with the first responder professional development environment which already covers a wide range of non-AV related topics on which they must also stay up-to-date. Examples of national training curricula which could be amended to include topics related to driving automation include the National EMS Education Standards and the Standard for Fire Service Professional Qualifications Accreditation and Certification Systems (NFPA 1000).

Case Study - PennDOT AV Incident Response Field Guide: PennDOT has developed its AV Incident Response Plan (AVIRP) in response to state legislation (Act 130) that legalizes the commercial operations of AVs. The AVIRP includes both a communications flowchart and a mobile application known as the AV Incident Response Field Guide (AVIR-FG). The AVIRP communications flowchart identifies the key stakeholders between state and local agencies involved in AV-related incident response, defines roles and responsibilities, and outlines communications avenues from AV Incident Detection to Post-Incident Activities. The AVIR-FG app as shown in Fig. 1 supports the implementation of the AVIRP with real-time communications and reporting on incidents involving an AV. Its goals are to 1) ensure all responders (law enforcement, fire, towers) are aware of appropriate incident response considerations to ensure they safely respond to an incident involving an AV or a personal delivery device (PDD); 2) ensure communication is maintained to assist with response and interagency coordination efforts; 3) ensure PennDOT is aware of all reportable AV/PDD incidents (as the lead regulatory agency for AVs in Pennsylvania).

Case Study - Waymo First Responder Guide: Waymo has developed a comprehensive First Responder Program and Guide to ensure the safe operation of its ride-hailing service, Waymo One. The Waymo First Responder Program has trained over 15,000 first responders, including police, fire, and emergency medical personnel from more than 75 agencies across the United States. The company also actively engages with key first responder associations, such as the International Associations of Chiefs of Police and Fire Chiefs. The Program was recently certified by TÜV SÜD as meeting industry standards for emergency response protocols, aligning with AVSC's Best Practice [6].

The Waymo First Responder Guide, first developed in 2018, ahead of the Waymo One launch, reflects collaborations with first responders and has been regularly updated to reflect changes in technology, operational design domain (ODD), and regulatory guidance. The guide is designed for trained first responders and assumes a professional-level

Fig. 1. AVIR-FG Application for AV Incident Reporting [5]

background in safely responding to emergencies, including those involving damaged vehicles. Information in the guide is organized from general orientation to more specific details on responding to various scenarios and hazards, covering topics such as identifying a Waymo vehicle (see Fig. 2 below), disabling the ADS, turning the vehicle off, and accessing vehicle data. It also includes a dedicated hotline for first responders to communicate directly with Waymo on a 24-h basis.

Fig. 2. Illustrative section from Waymo's first responder guide [7].

2.2.3 First Responders Interacting with AVs in Traffic

In the breakout group focused on AVs interacting with first responders in traffic, the group sentiment was mostly negative in terms of the current situation, although some possible ways to improve these interactions were discussed, including the general improvement of ADS capabilities as the technology matures. In fact, the only "rose" that was mentioned was an anecdotal incident with a robotaxi in Phoenix in which the robotaxi did pull over correctly as a first responder with lights flashing was overtaking from behind.

There were several "thorns" discussed by the group. In particular, it was noted that not all AVs have the same capability and behavior, and so this makes the interactions with first responders unpredictable and often less than ideal. The group also discussed how human traffic controllers (HTCs) have gestures and equipment that are not standardized. In the case of human drivers, this lack of standardization generally does not cause issues because humans can extrapolate and understand new gestures with unfamiliar equipment. However, ADSes currently on the road have much more difficulty in such a situation. The conundrum of what AVs should do in the case of ad hoc HTCs (and particularly in the case of the "Bad Samaritan") was also discussed as an on-going issue. The issue of calls to public safety answering points (PSAPs) being normally tied to a jurisdiction was mentioned for AV services that cross jurisdictional boundaries, as well as the variability of both the information being supplied to the PSAPs and the processes of the PSAPs themselves. More standardization in this area was highlighted as a needed development.

For "buds", the group had a disparate set of ideas that were discussed. Digital infrastructure in the form of sensors, edge/cloud compute for situational awareness, and communication (specifically machine-readable) could all assist with first responder/AV interactions. It was noted that some jurisdictions collaborate and coordinate their PSAPs across jurisdictions, with Maricopa County in Arizona being a prime example of the benefits of this collaboration. The group noted that more effort could be put into collaboration across larger jurisdictions and standardization of PSAP practices. "Stay away orders" or geofences to help AVs avoid areas with an ongoing incident were seen as a promising technique for reducing the number of AV-first responder interactions in traffic. San Francisco, Phoenix, and Austin have all started to implement geofences.

The group discussed the possibility of the AV displaying some sort of standardized light or signal that indicates whether the ADS is engaged or not as well as its intentions. One effort discussed was SAE Recommended Practice J3134 on ADS Marker Lamps [8]. This standard has known limitations such as the restriction of the lamp to viewing from the front of the vehicle and the decision to defer signals of the ADS intentions to future versions, however, group members noted that the standard is in the process of being updated. While not related to first responder interactions with AVs in traffic, the group discussed the usefulness of a heads-up display in the vehicle. This could provide information for first responders on the scene of an incident where the AV is stopped, as well give passengers improved situational awareness.

The final bud discussed was the in-progress Recommended Practice from SAE, SAE J0911, on first responder interactions with fleet-operated ADS-dedicated vehicles (ADS-DVs). This Recommended Practice is based on AVSC's best practice and is expected to be published in 2025. The SAE J0911 document effort is inclusive of the perspectives of infrastructure owner-operators (IOOs) and first responders to produce guidance that is useful for various stakeholders and ensures that additional, undue burdens are not placed on first responders who already have many other responsibilities.

2.3 Data Sharing and Standardization

A major premise for this panel was the need for standardized and accurate first responder interaction and crash data sharing that identifies the various levels of driving automation. Good data is essential for the traffic safety community to conduct better data analysis and targeted education. There was much discussion on the characteristics of quality, accurate and uniform data of AV crashes and interactions, the importance of understanding the data, and ideas for standardization of identifying automated vehicles.

Both ADS developers and government agencies can benefit from good data that can provide a consistency in learning and understanding the logic and anomalies, understanding operational responsibility, and to consider appropriate best practices, regulations, and policies. If there is a level of comfort on agreed-upon data sharing, this can potentially demonstrate ADS safety performance which can lead to higher acceptance levels – and help with scaling.

If crash data on AVs can be standardized and used appropriately, it can be used like other data collected now for traffic safety analysis, such as crash analysis based on vehicle size, seat belt use, and demographic and situational information. This analysis can be used for AVs to determine ways to reduce crashes through targeted education as well as for developing a strong understanding of design capabilities and limitations.

To be assured of data quality, the panel discussed a desire for properly vetted and standardized ways to identify the levels of driving automation, assurance that all developers identify each vehicle consistently, and the provision of this information transparently to appropriate sources such as motor vehicle administrations. SAE's levels of driving automation, defined in Recommended Practice J3016 [9], have been widely adopted, however, attempts to integrate these levels with registration and crash data have not led to these being reliable sources of that information.

As an example, Maryland has incorporated SAE level fields in registration data since 2022 and crash data since January 2024. Out of over six million registered vehicles, only 152 are designated with an SAE level other than 0 (no driving automation). There are two vehicles registered as SAE level 5, and the vast majority of non-level 0 vehicles are described as level 1 or 2, even if they are too old to have any driver support features. From the crash data, over 1,400 vehicles were reported as level 4 or 5 out of over 92,000 vehicles involved in crashes in the first six months of 2024, and some vehicles (e.g., Teslas) which likely have driving automation systems installed (even if not engaged at the time of a crash) are often categorized as level 0.

While there is no current way to populate the field from reliable, standardized sources, the Model Minimum Uniform Crash Criteria (MMUCC) 5th edition recommends including a data element to collect the SAE level for all vehicles and level of automation engaged at the time of the crash. Due to the difficulties in officers collecting this information at the scene, in the 6th edition of the MMUCC, the recommendation has been changed to only collect the information when it is known [10].

Implementation of consistent SAE level identification from a trusted source can provide ease of implementation and uniformity across jurisdictions via shared official databases with state titling and registration systems. Title and registration designation can be available to licensing agencies, insurance agencies, courts, law enforcement, and crash data to provide consistency throughout databases.

3 Conclusions and Next Steps

A key takeaway from the session is that first responders must be part of the dialogue when it comes to training and interaction procedures. Opportunities to engage first responders exist both prior to testing and deployment being planned in a jurisdiction and while these activities are ongoing. A structured approach can help to respect the limited time emergency professionals can dedicate to these efforts.

One set of principles to build on is the American Association of Motor Vehicle Administrators (AAMVA) Guidelines for Regulating Vehicles with Automated Driving Systems [11]. This document recommends that States and provinces include processes to encourage collaborative efforts with first responders. AAMVA recommends that each state have a permit or approval process for ADS testing and deployment in order to provide the reviews and structure. AAMVA also recommends that states develop law enforcement protocols, with input from ADS developers, and that each developer creates a law enforcement interaction plan for their specific vehicle models. While the AAMVA guidelines currently only refer to law enforcement, there is recognition of the importance of including other first responders.

States or metropolitan regions may also wish to designate points of contact to coordinate conversations with leadership from first responder organizations, maintain a list of jurisdictions and their primary emergency response points-of-contact, and facilitate opportunities for training and demonstrations to allow responders to experience an ADS in action.

On the topic of data standardization, future research could explore processes and criteria to ensure the proper level of detail and trusted sources for registration and crash reporting. Considerations include whether SAE levels are the most appropriate identification for registration, whether information could be encoded into a VIN or displayed on a manufacturer label, and the effects of aftermarket products and over-the-air updates.

Moving to more on-the-ground issues, research can advance the "buds" discussed in the breakout groups to improve communications and interactions at a scene. As one example, several regions have seen success with geofencing to reduce the number of interactions first responders and ADS-equipped vehicles have. However, more work can be done to refine the details – how large the geofence should be, how long it should last, understanding what happened when it does not work as planned – and to standardize approaches and data transfer methods if appropriate.

Future research efforts should leverage the efforts of the organizations who have developed the documents referenced in this chapter – for example, AVSC, SAE, GHSA, and AAMVA – along with additional stakeholders that should be brought into the discussion, for example, the National League of Cities or the International City/County Management Association. Additionally, research should build on related projects such as a recent National Cooperative Highway Research Program project on the impacts of connected and automated vehicle technologies on traffic incident management response [12].

Given the collection of work documented here, which dates back to 2018, and the maturity of the practice of managing interactions between AVs and First Responders, a national dialogue series could be created centered on First Responder Interactions with AVs. With support, the efforts expended to date can be constructed into the foundation of a national community of practice. The series would act as a method and continuum for collecting, harmonizing and disseminating experiences and best practices while also informing others yet to have experienced AVs in their community.

References

1. Terry, T., Trimble, T.E., Blanco, M., Fitzgerald, K.E., Fitchett, V.L., Chaka, M.: An Examination of Emergency Response Scenarios for ADS. Crash Avoidance Metrics Partners LLC, Farmington Hills, MI (2018). https://www.vtti.vt.edu/PDFs/Final%20Report%20-%20An%20Examination%20of%20Emergency%20Response%20Scenarios%20for%20ADS.pdf
2. Trimble, T., Terry, T.: Law enforcement, first responder and crash investigation preparation for automated vehicle technology. Virginia Tech Transportation Institute (VTTI). Prepared for Governors Highway Safety Association and State Farm (2021). https://www.ghsa.org/resources/Preparation-for-Automated-Vehicle-Technology21
3. Automated Vehicle Safety Consortium: Revision of Best Practice for First Responder Interactions with Fleet-Managed Automated Driving System-Dedicated Vehicles (ADS-DVs). SAE Industry Technologies Consortium. AVSC-I-01-2024 (2024). https://www.sae.org/standards/content/avsc-i-01-2024/
4. Goodison, S.E., Barnum, J.D., Vermeer, M.J.D., Woods, D., Lloyd-Dotta, T., Jackson, B.A.: Autonomous road vehicles and law enforcement: identifying high-priority needs for law enforcement interactions with autonomous vehicles within the next five years. RAND Corporation (2020). https://www.rand.org/pubs/research_reports/RRA108-4.html
5. Rhone, G.: Preparing for the AV Bill. Penn State Transportation Engineering and Safety Conference (2022). https://www.tesc.psu.edu/assets/Session%20PDFS/TESC%20Session%20Presentations/PennDOT--Moving-ahead-with-Emerging-Mobility-Preparing-for-AV-Bill.pdf
6. Waymo: Waymo's First Responder Program Receives Independent Safety Confirmation (2024). https://waymo.com/blog/2024/12/waymos-first-responder-program-receives-independent-safety-confirmation
7. Waymo: Jaguar I-PACE Waymo Emergency Response Guide and Law Enforcement Interaction Protocol (2024). https://storage.googleapis.com/waymo-uploads/files/documents/first-responders/Emergency%20response%20guide%20and%20law%20enforcement%20interaction%20protocol%20.pdf
8. SAE International: Surface Vehicle Recommended Practice: Automated Driving System (ADS) Marker Lamp. SAE J3134 (2019). https://www.sae.org/standards/content/j3134_201905/
9. SAE International: Surface Vehicle Recommended Practice: Taxonomy and Definitions for Terms Related to Driving Automation Systems for On-Road Motor Vehicles. SAE J3016 (2021). https://www.sae.org/standards/content/j3016_202104/
10. National Highway Traffic Safety Administration: MMUCC Guideline: Model Minimum Uniform Crash Criteria, 6th Edition. DOT HS 813 525 (2024). https://crashstats.nhtsa.dot.gov/Api/Public/ViewPublication/813525
11. American Association of Motor Vehicle Administrators: Guidelines for Regulating Vehicles with Automated Driving Systems, Edition 4 (2024). https://www.aamva.org/getmedia/c95fd480-5917-471a-b7ee-da88ec6cb7b5/Guidelines-for-Regulating-Vehicles-with-Automated-Driving-Systems-Ed-4_final.pdf
12. Son, P., et al.: Impacts of connected, automated vehicle technologies on traffic incident management response. national academies of sciences, engineering, and medicine. NCHRP Web-Only Document **395** (2024). https://doi.org/10.17226/27655

Part II: Business Models and Operations

Entrepreneurship for Automation: Growing Sustainable Business Models for Automated and Autonomous Transport Worldwide

William Riggs[1](✉), Henriette Cornet[2], Tom Alkim[3], Ingrid Skogsmo[4], Manabu Umeda[5], Matthew Daus[6], and Arnd Bätzner[7]

[1] University of San Francisco, San Francisco, USA
wriggs@usfca.edu
[2] Urban Innovate / University of San Francisco, San Francisco, USA
hcornet@usfca.edu
[3] MapTM, Utrecht, Netherlands
tom.alkim@maptm.nl
[4] Swedish National Road and Transport Research Institute (VTI), Gothenburg, Sweden
ingrid.skogsmo@vti.se
[5] University of Tokyo, Tokyo, Japan
umeda@iis.u-tokyo.ac.jp
[6] IATR/Windels Marx, New York, USA
mdaus@windelsmarx.com
[7] Baetzner Metropolitan, Zurich, Switzerland
arnd@baetzner.ch

Abstract. The advent of automated and autonomous vehicle (AV) technology heralds a transformative shift in transportation, with profound implications for sustainability and urban mobility. This paper examines the transference of private sector innovations to public sector applications to foster sustainable AV business models around the world. Through collaboration with 30 experts, entrepreneurial business models were developed, tested and refined using a criteria matrix—centering around two clustered service concepts. Through rounds of feedback, dominant themes emerged. The findings underscore the potential for autonomous vehicles to enhance urban mobility, reduce emissions, and improve operational efficiencies, but areas where private and business innovation can provide innovative entrepreneurial solutions for public transport that pave the way for more resilient and sustainable transportation systems.

Keywords: autonomous vehicles · entrepreneurship · business models · public transport · transit

1 Introduction

As automated and autonomous vehicle (AV) technology has evolved, there are increasing efforts to translate private innovations to public sector models of sustainable transit. Use cases for shared rides, shared vehicles and lightweight transit platforms have been

G. Meyer and S. Beiker (Eds.): ARTSymposium 2024, LNMOB, pp. 41–62, 2026.
https://doi.org/10.1007/978-3-032-12374-9_4

introduced across many countries [1–3]. This paper explores use cases emerging from expert panels conducted in 2024 and evaluates different business models in which AVs are being introduced to complement and supplement resilient transit systems worldwide [4–6].

The paper acknowledges the marketplace for logistics in the AV ecosystem, but specifically focuses on emergent models relating to autonomous vehicle operations that relate to ridesharing use cases that carry passengers including: 1) **On-demand + Door-to-door or On-demand + Flexible Pick-Up and Drop-Off areas (PUDOs)** (for example lightweight AV rideshare / robotaxis); and 2) **Scheduled + Virtual stops or Scheduled + Fixed stops** (for example public-transit-oriented shuttles). These models can be broadly defined based on the descriptions in the bullets below.

1. On-demand + Door-to-door or On-demand + Flexible Pick-Up and Drop-Off areas (PUDOs): This business model is similar to traditional ridesharing service with door-to-door or fixed service. It involves providing transportation services for passengers on a per-ride basis, similar to traditional taxi services. Under this model an AV could be used to provide transportation for customers who need a vehicle for a short period of time; a fleet that is procured for a certain time per day, with customers paying for the time that they use the vehicle. This would be similar to the model employed by companies like Uber and Lyft, who have been experimenting with AV technology and partnerships, and could ultimately transition to fully autonomous fleets.
2. Scheduled + Virtual stops or Scheduled + Fixed stops: Largely predicted to be some of the earliest and most popular early deployments due to capital and policy factors, this model aligned most with autonomous public transit with most services providing transportation for passengers along predefined routes, with either undesignated or designated stops. This model involves developing autonomous buses or shuttles that can transport passengers to and from transit hubs or other key locations within a city or region. This can be in the form of paratransit or wheelchair accessible vehicles (WAVs) and also be complementary with existing high-capacity transit corridors

This paper specifically dives into entrepreneurial futures as opposed to regulatory frameworks, and therefore does not focus on the unique regulatory environments that AV deployments reside in. This is consistent with more recent views on entrepreneurial activities [7, 8] which account for a broader view of value creation that takes into account the public good and factors related to ESG (environment, social and governance) goals.

2 Background

The deployment of autonomous vehicles presents vast opportunities. The increased efficiency from AVs could save time and money for both individuals and businesses. It is linked with good land use planning and the principles of "Transit Oriented Development" (TOD), calling for the development of new or densification of existing residential and commercial real estate primarily in the vicinity of transit stations, promoting intermodality and enhancing walkable access to transit, which in a long-term perspective has shown to lower direct costs for parties involved and minimize negative externalities [9].

In this light AV rideshare services complementing fixed-route transit systems can fit into the logic of TOD, acting as tools in an intermodal travel chain toolbox. They also have the potential to help achieve greater service quality and lower service delivery costs (e.g. in currently non-TOD compliant areas where today's low-frequency bus services fail to provide an offer attractive to many of today's private car users) [10].

Coupled with this broad increase in accessibility, AVs have the potential to improve mobility for people who are unable to drive, such as the elderly or disabled, and to vastly reduce the number of accidents caused by human error, saving thousands of lives each year. This increased accessibility and safety could improve their quality of life broadly and contribute to economic growth through reduction of insurance risks and the cost of accidents. That said, as with any new technology, it will be important to address the tensions and maximize the opportunities to create a more equitable, sustainable, and efficient transportation system. Their vehicles could lead to the development of new business models and services, such as ridesharing, logistics, and delivery services. These new models could create new opportunities for entrepreneurs and businesses.

And these opportunities are global in nature, making collaboration essential [11]. In the US alone, the Alliance for Automotive Innovation has recognized 84 Autonomous Vehicle companies operating across 30 states and 120 cities in the USA. In the US, the most popular form factor currently in use are a rideshare model largely focused in San Francisco, Phoenix, and Las Vegas [12]. In locations like Europe and Asia preliminary use cases centered around fixed route demonstration projects, usually embedded in public transit systems or on isolated campuses, however there are increasing examples from companies including VW, WeRide and Holon where robotaxi deployments are also integrated into the transportation system.

In the US, the companies that have become most widely known for AV ride hailing service include Cruise (a GM subsidiary suspended late 2024) who are majority owned by General Motors (GM) and Alphabet owned Waymo both of which have demonstrated level-4 autonomous service for public riders. While these companies have yet to enter other markets there are increasing entries into the market in Europe and Asia that create a robust competitive landscape for future AV business models.

2.1 Case Studies in the United States

As stated previously, the US may have an edge in terms of public, level-4 AV deployments, followed closely by many fertile deployments in China by the likes of Baidu, Pony.ai and WeRide. For example from 2022-late 2023 Cruise, LLC received all required permits for autonomous driving and successfully launched fared rides to clients within a restricted operational design domain (ODD) in San Francisco [13] [14]. In addition, they launched fared autonomous service for smaller ODDs in Phoenix and Austin in 2022 [13].

All of these operations used a modified GM Chevrolet Bolt, however at the same time the company was testing a "production version" of an autonomous shuttle, the Cruise "Origin" which was designed as a multi-passenger AV with seats that face each other, sliding doors and no driver controls [13, 14]. While work on this platform was suspended in 2024 [15] it represented a business model that focused on developing two kinds of vehicle platforms—both smaller, last-mile robotaxi vehicles as well as larger, shared AV shuttles.

During a similar timeframe, Waymo received the necessary approvals to provide driverless, fared rides in San Francisco and Phoenix [16], expanding to the Los Angeles market in 2024 [16–18]. They have charted out plans to partner with Uber and pursue profitability scaling to 10 cities within 2025–2026 horizon [19–21]. At the same time other companies have continued to scale their products, through investment, pilots and strategic partnerships; for example Aurora (AV logistics and software development), Zoox (AV shuttle and software development), and Nuro (AV logistics and last mile delivery service) to name a few [22].

In addition to these private sector companies developing technology, public organizations such as Jacksonville Transportation Authority (JTA), the San Francisco County Transportation Authority (SFCTA), and Michigan State University (MSU) have also conducted AV deployments, but these have been less technologically advanced than the corresponding US private sector deployments [17]. In Jacksonville an AV shuttle project envisions a 10-mile elevated and automated shuttle service, in part by repurposing the current 33-year-old monorail system [23]. The project includes a 3-mile autonomous shuttle line that will assist the existing monorail to connect major sports stadiums and entertainment venues in downtown Jacksonville [24]. The $49M project was funded by a $12.5M US Department of Transportation BUILD grant and funding from the Florida Department of Transportation, the North Florida Transportation Planning Organization and JTA and involves connected infrastructure such as cameras, antennas, lidar and radar sensors at key intersections in the corridor with the goals of better connecting downtown neighborhoods.

The Michigan AV Bus project has been operating since early 2022 with the 27-foot, 22 seat, fully electric and level 4 autonomous Karson ATAK bus providing rides to students, faculty, staff and visitors on a 2.5-mile route across the MSU campus [23]. The aim of this project, between the State of Michigan, bus manufacturer Karsan, and Michigan-based AV software development company ADASTEC, was to allow riders to get a glimpse of AV technology. Similarly, the company May Mobility ran a free AV shuttle service at a farmers market in Ann Arbor from around 2021 [25]. This included 4 hybrid Lexus SUVs and 1 wheelchair accessible vehicle, driving in a circular route of 21 designated stops with a supervising AV safety driver [25].

The main aim of the May Mobility project was again to experiment with AV technology while integrating it with public transit and decreasing demand for personal vehicles, and SFCTA had similar goals in a demonstration pilot during 2023 [26]. The project provides a dedicated loop of transit along fixed stops on Treasure Island in San Francisco Bay, providing service to residents to connect to local transit. While there was ample interest in the deployment's connectivity with public transit, there were stated challenges with the jarring performance of the vehicle technology, with one journalist describing how, "San Francisco's first driverless shuttle somehow manages to make whiplash-inducing stops despite never exceeding a few-miles-per-hour crawl." [27].

At the same time, despite challenges, this technological experimentation has not gone unnoticed globally. For example, in the context of the 2023 Automated Road Transportation Symposium (ARTS) and the International Task Force on Vehicle Highway Automation (ITFVHA), which both took place in San Francisco in July 2023, more than 20 European experts from 9 different member states took the opportunity to ride with Cruise and Waymo [28]. They were interviewed about their experiences gathered in the 86 rides in total, an average of 4.3 rides per person. For 80% of the respondents, this was the first ever experience in a commercially deployed self-driving vehicle, not counting the many test rides most of them had with prototypes and safety drivers under controlled conditions.

Over 90% of the respondents indicated that this experience had changed their perspective on the work they are doing. The most cited positive experiences were related to the driving style of the vehicles which was smooth and assertive. Interactions with other traffic, including detecting and stopping for pedestrians, were better than expected. Riders noticed that a self-driving vehicle was patient and obeyed traffic rules—even when other adjacent road users perceive AVs as "too slow" (sticking to the speed limit) or are coming to a complete stop at every stop sign (instead of a "rolling" stop). They identified potential challenges to pick-up and drop-off zones, sometimes erratic driving behavior, and interactions with emergency vehicles, and potential interruption of traffic flow as lessons that could be learned or mitigated as the technology matures and expands beyond the US market.

2.2 Case Studies in Europe

Transitioning to Europe there has been strong interest in learning and collaboration particularly with the United States (US) and Japan—with numerous international workshops, a dedicated project called Framework for coordination of Automated Mobility in Europe (FAME), and dedicated AV topics among the clusters of the European Partnership on Connected, Cooperative and Automated Mobility (CCAM Partnership). Reports have outlined cooperation opportunities such as [29]:

- US initiatives like the Infrastructure Investment and Jobs Act of 2021, providing collaboration opportunities in areas such as decarbonization and traffic management systems.
- South Korea's commitment to SAE Level 4 Field Operational Testing (FOT) opening avenues for joint ventures in autonomous vehicles.
- United Kingdom's safety and legal frameworks as well as research and development funding leveraging private sector investment.
- Switzerland's SAAM (Swiss Association for Autonomous Mobility) focus on shared automated mobility solutions, offering collaboration prospects in public transportation and the development of a comprehensive Transport as a Service (TaaS) ecosystem.
- Australian exploration of opportunities in safety benchmarking and mass transit feasibility studies.
- Canadian involvement safety assessments and infrastructure development, aligning with global efforts in CCAM.
- Dubai's ambitious self-driving strategy and a strong focus on smart Traffic Management, road infrastructure enhancement, and shared autonomous transportation options.
- China's approach to a growing OEM sector, technological innovation and intelligent infrastructure development.

- Israel, as a thriving technology hub and opportunity for collaboration in technology development and regulatory alignment.

In this light many countries in Europe have begun to experiment with autonomous transportation models. While regulation and compliance of AV hardware and software have posed a challenge the EU has been a leader in the of the fixed route AV transport space, largely due to the work of French companies Navya and EasyMile who have both completed over 100 tests across multiple countries and cities [30]. While the high costs to scale large-scale AV fleet operations, have posed a challenge to these deployments, the entry of European automotive suppliers like Bosch, Continental, and ZF into the space could create more competition and accelerated product development in the AV segment in the imminent future.

One of the strongest features of these European deployments, and how they have aligned with CCAM concepts, has been their use of technological advancements to realize the potential for increased safety, reduced environmental impacts, and inclusiveness. For example, *Ride the Future* (RTF) was a collaboration project between seven partners[1] in Linköping, Sweden,[2] deploying electrified, automated shuttles as a new mobility solution complementing public transport [31]. The operation of shuttles (originally Navya DL4 but replaced by 2 EasyMile EZ10 Gen2) started in spring 2020; from autumn 2024 the operation transitioned to on-demand traffic in an extended area with over 50 bus stops, to provide a service for all end users, including children with special needs, elderly, other groups with specific travel needs, such as families. The speed limit for the project was between 30 and 40 km/h; however, the Operational Design Domain (ODD) of the automated shuttles' speed was lower than 20 km/h. At the time of the deployment in Sweden, there were no approved operations of shuttles without a safety driver onboard, thus in this project the safety driver had the same responsibility as a normal bus driver.

In the beginning, the focus has been on vehicle performance and technical issues, such as solving brake behavior, positioning, signage, but many challenges persisted, such as developing business models, legal issues or perceived and actual safety for passengers. Important learnings included the key point that—running an operational business that is accessible to a high enough degree for citizens to see it as a real mobility alternative is challenging—and it generally takes time. This is particularly the case if this is a new transit service.

Conversely some European concepts for AV fleet deployment as part of existing public transit thus specifically address planning challenges such as structural regional car

[1] The partners include Akademiska Hus, Linköping Municipality, Linköping University, RISE – Research Institutes of Sweden, Swedish National Road and Transport Research Institute (VTI), Transdev Sweden and Östgötatrafiken.

[2] Linkoping is the fifth largest city in Sweden with a growing population of 167,000 inhabitants (in 2024). The site includes both a mixed traffic with separate lanes for Vulnerable Road Users (VRUs) and shared spaces with VRUs. The Area of operation included Vallastaden, a newly built residential area, and Linköping University's Campus. It should be noted that Vallastaden was built with sustainability in mind; with relatively few parking spaces and is optimized for walking and cycling; location also includes school for children with and without special needs and a retirement home for elderly people with cognitive disabilities).

dependency in suburban spaces: A 5-year project in Greater Zurich region, Switzerland starting operations mid-2025 specifically focuses on the exurban Furttal Valley region with a topography that limits the efficiency of conventional bus services [32]: An AV fleet will be tested as per its capacity to efficiently complement the valley's central rail corridor, avoiding service parallelisms with fixed-route transit. A key objective in the Furttal Valley trial is better insights into requirements of future governance frameworks for automated road transit services, econometrics of service deliveries, operational aspects and cost structures.

2.3 Case Studies in Asia

Asia provides similar case studies. For example, in Japan, which is facing a declining birthrate and an ageing society, the government has set a target of deploying around 50 locations where regionally limited driverless automated mobility services are provided by FY2025, and around 100 locations by FY2027, and is actively promoting a realization of automated mobility services into the society.

A permission system for specified automated operation, which is driverless automated driving, equivalent to Level 4 in the SAE (Society of Automotive Engineers) Levels of Driving Automation, was established by the Act Partially Amending the Road Traffic Act, and the provisions on automated driving took effect in April 2023 [33]). In May 2023, Machidukuri ZEN Connect Inc. Started Japan's first Level 4 automated driving mobility service operation in Eiheiji Town, Fukui Prefecture [34]. And in June 2024, Kajima Corporation and BOLDLY Inc. Were granted permission to operate buses under Level 4 automated driving in Haneda Innovation City, Tokyo [35].

In FY2024, the Ministry of Land, Infrastructure, Transport and Tourism, Logistics and Road Transport Bureau selected around 100 locations for FOTs to support automated mobility service deployment, and many FOTs have been conducting nationwide in Japan. However, most of these projects rely on government subsidies, and how to operate automated driving services in a sustainable manner is a major issue in Japan.

Similarly, China has continued to make strong progress in the autonomous vehicle space largely due to ongoing advancements in both technological innovation and policy support [36]. China's booming auto manufacturing industry has created a solid foundation for many AV companies to settle and build their operations in the country. Mo Luyi, Vice President of AV producer Pony.AI has also recognized China's uniquely complex traffic conditions as an opportunity to collect more valuable data, which he believes will be helpful for technological upgrades.

Following testing of approximately 20 million miles between Beijing, Shanghai, Guangzhou, and Shenzhen, tech giant Baidu received permission to launch their AV rideshare service on public roads in Wuhan and Chongqing. Baidu officially launched publicly in August 2022 with their AVs, powered by their AV technology called Apollo Go, operating in designated areas in both cities. Baidu has announced that they plan to expand their ride-hailing AV services to 65 cities by 2025 and 100 cities by 2030 [13].

Another Chinese company that is currently actively involved in implementing AV technology across many different operational models is WeRide. Headquartered in Guangzhou a city recognized as an avid promoter of innovation and entrepreneurship. WeRide's work is focused on the AV models of transportation, freight, and sanitation

[36]. WeRide established its driverless sanitation vehicle "Robosweeper" on public roads in the Nansha District as early as May 2022 with as many as 50 vehicles on the road [37]. The key incentives for the use of AV sanitation vehicles like the "Robosweeper" are to leverage AV technology to provide environmental services such as sweeping, sprinkling and disinfectant spraying to efficiently keep cities safer and greener, without the need for any human contact.

Likewise Shenzhen-based DeepRoute.ai has recognized the challenge of high costs in producing AV technology. They have successfully managed to decrease the cost to equip their AV rideshare service and medium-duty trucks with Level-4 AV infrastructure to less than $10,000 US per vehicle. Interestingly, Zhou Guang – the founder of DeepRoute.ai has announced that they plan to continue to investigate ways to integrate their systems into consumer vehicles, with the aim being to provide consumers access to an affordable Level-4 AV solution [36].

2.4 Entrepreneurial KPIs

Within the scope of this global innovation, there is a growing need for entrepreneurship and business model refinement. While each of the global examples mentioned operate in a varied regulatory frameworks, with activities as brokered cooperatively between different kinds of government agencies, industry stakeholders, and other parties, each case also has distinct opportunities and key performance indicators that play a core role in responding to regulatory concerns while creating value in the market, gaining consumer traction and generating money over invested capital (MOIC) for investors.

In the example of Cruise and Waymo, two of the companies evaluated in the deployment of AV rideshare services, companies have established key performance indicators (KPIs) to measure the success of their AV rideshare services. Here are some examples of the KPIs for each company:

- Cruise:
    - Safety: Cruise had set a goal of achieving zero accidents or incidents involving its autonomous vehicles. The company was tracking safety-related metrics, such as the number of disengagements (when the human safety driver takes control of the vehicle), the number of accidents and incidents, and the frequency of safety-critical events.
    - Passenger Experience: Cruise was also focused on providing positive passenger experience. The company was measuring metrics such as passenger satisfaction, waiting times, and vehicle cleanliness.
    - Utilization: Cruise aimed to maximize the utilization of its vehicles to ensure efficient and profitable operations. The company was tracking metrics such as the number of miles driven per day, the percentage of time vehicles are in use, and the number of trips per day.
- Waymo:
    - Safety: Waymo has set a similar goal of achieving zero accidents or incidents involving its autonomous vehicles. The company is tracking safety-related metrics,

such as the number of disengagements, the number of accidents and incidents, and the frequency of safety-critical events.
- Rider Experience: Waymo is also focused on providing a positive rider experience. The company is measuring metrics such as rider satisfaction, wait times, and vehicle cleanliness.
- Efficiency: Waymo aims to optimize the efficiency of its operations to maximize profitability. The company is tracking metrics such as the number of miles driven per day, the percentage of time vehicles are in use, and the number of trips per day.
- Sustainability: Waymo is committed to reducing the environmental impact of its operations. The company is tracking metrics such as the number of miles driven on electric power, the number of passengers transported per mile, and the reduction in greenhouse gas emissions compared to traditional vehicles.

Overall, both Cruise and Waymo are focused on safety, passenger/rider experience, efficiency, and sustainability, among other factors in addition to their pursuit of profitable AV rideshare businesses. These KPIs are critical to the success of their services but also relate to their intention to pursue varied form factors into the future as they help the companies ensure that they are providing a safe, reliable, and cost-effective transportation option for their customers. Other companies that offer different operating platforms seeking to develop and deploy autonomous shuttle services have similar yet varying KPIs. For example, the companies Navya and EasyMile, which have operated as a part of many EU projects, specialize in the development and deployment of autonomous shuttle services. Both companies have established key performance indicators (KPIs) to measure the success of their services. They include:

- Navya:

  - Safety: Navya places a strong emphasis on safety and has set a goal of achieving zero accidents or incidents involving its autonomous shuttles. The company tracks safety-related metrics, such as the number of disengagements, the number of accidents and incidents, and the frequency of safety-critical events.
  - Passenger Experience: Navya is focused on providing a high-quality passenger experience. The company measures metrics such as passenger satisfaction, wait times, and vehicle cleanliness.
  - Efficiency: Navya aims to optimize the efficiency of its operations to maximize profitability. The company tracks metrics such as the number of miles driven per day, the percentage of time vehicles are in use, and the number of trips per day.

- EasyMile:

  - Safety: EasyMile is also focused on safety and has set a goal of achieving zero accidents or incidents involving its autonomous shuttles. The company tracks safety-related metrics, such as the number of disengagements, the number of accidents and incidents, and the frequency of safety-critical events.

- Passenger Experience: EasyMile aims to provide a positive passenger experience. The company tracks metrics such as passenger satisfaction, wait times, and vehicle cleanliness.
- Accessibility: EasyMile is committed to providing accessible transportation for all passengers, regardless of their mobility needs. The company tracks metrics such as the number of wheelchair users transported, the availability of accessible vehicles, and the ease of boarding and exiting the vehicle.
- Environmental Sustainability: EasyMile is committed to reducing the environmental impact of its operations. The company tracks metrics such as the number of miles driven on electric power, the reduction in greenhouse gas emissions compared to traditional vehicles, and the use of sustainable materials in vehicle manufacturing.

In sum, complementing the attributes of companies like Waymo and Cruise, shuttle manufacturers like Navya and EasyMile are focused on safety, passenger experience, efficiency, and sustainability, among other factors—despite focusing on more public-transit-oriented deployments. This overlap is critical in developing sustainable business cases for AVs—particularly if they are to be integrated within public transportation systems, either as a supplement to or complement to or replacement of existing public transport.

3 Methodology

A structured approach using expert surveys and a criteria matrix was used to develop and refine business models for autonomous vehicle (AV) services. Thirty-three (33) experts with substantial experience in transportation and business models were used as a part of a workshop at the Automated Road Transport Symposium 2024 in San Diego. These experts included professionals from academia, industry, and government sectors. As defined in Table 1.

A series of idea development surveys were designed to gather insights on the viability and effectiveness of various AV service models. Questions were structured to explore opinions on On-demand and Scheduled service models, specifically: 1.On-demand + Door-to-door or On-demand + Flexible Pick-Up and Drop-Off areas (PUDOs); 2.Scheduled + Virtual stops or Scheduled + Fixed stops (e.g., public-transit-oriented shuttles).

The surveys were distributed electronically, ensuring ease of access and response. Participants were asked to rate each business model based on several criteria and provide qualitative feedback on potential challenges and opportunities around 3 core questions:

- Imagine and describe the location where the scenario would be deployed, if a user group is targeted and how it might operate. Where is it, who does it serve and how does it work?
- Which stakeholders are needed to make it work and who is in lead?
- What are the first challenges that you can think of that could hinder your deployment?

Table 1. Stakeholders

Expert Stakeholder Group	*Count*
Research: Other	1
Advocacy: Educational	1
Government: Public Transport Operator	2
Government: Other	2
Private: Other	2
Other	2
Government: Local / City	3
Private: AV Manufacturers	4
Private: Planning, Design, & Engineering Consulting	4
Research: Academic Institution	4
Government: National	8
	33

To refine the proposed business models, the responses from the surveys were analyzed using a criteria matrix. The criteria were chosen to ensure a comprehensive evaluation of each model's feasibility and impact. The criteria included:

- High Impact:
  - Evaluation of the potential of each model to significantly enhance urban mobility, reduce emissions, and improve operational efficiencies.
- Feasibility:
  - Assessment of the organizational, political, and technological feasibility of implementing each model. This involved considering the current regulatory environment, public acceptance, and technological readiness.
- Cost-Effectiveness:
  - Analysis of the economic viability of each model, including initial investment costs, operational costs, and potential revenue streams.
- Timeliness:
  - Evaluation of the ease and speed of implementation for each model. Factors such as infrastructure requirements, regulatory approvals, and stakeholder readiness were considered.
- Risk:

- Assessment of the risks associated with each model, including technological, operational, and financial risks. Models were rated based on their risk levels, categorized as low, moderate, or high.

- Sustainability:

  - Examination of how well each model aligns with Environmental, Social, and Governance (ESG) goals. This included evaluating the environmental benefits, social equity implications, and governance frameworks supporting each model.

Responses from the expert surveys were collected and quantitatively analyzed using the criteria matrix. Based on the analysis, the business models were refined to address identified weaknesses and enhance strengths. This iterative process involved re-evaluating models against the criteria matrix and incorporating expert feedback. The refined models were ranked based on their overall scores in the criteria matrix. The top-performing models were selected for further development and potential pilot implementation.

4 Results

Entrepreneurial case studies were first clustered into two consolidated areas, based on on-demand vs. scheduled service models: Cluster 1: On-demand + Door-to-door and On-demand + Flexible Pick-Up and Drop-Off areas; and Cluster 2: Scheduled + Virtual stops and Scheduled + Fixed stops, bus stops. These were then analyzed by the trends in entrepreneurial service models as shown in Fig. 1.

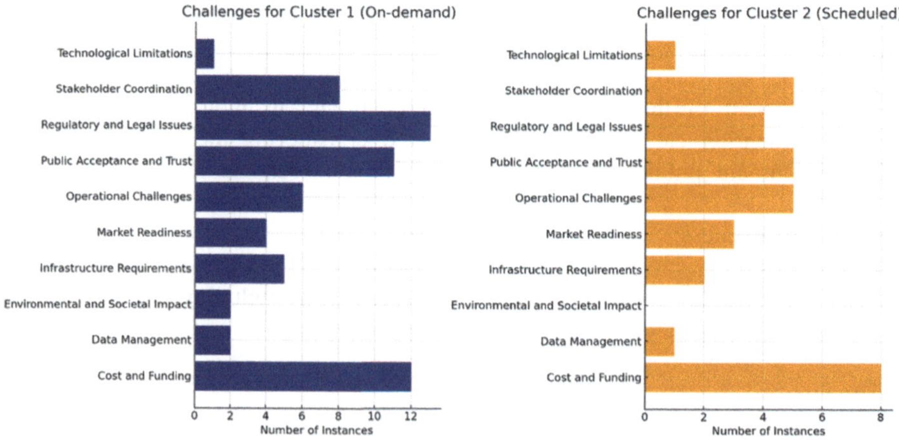

Fig. 1. Challenges by Entrepreneurial Cluster

For On-demand Services (Door-to-door, Pick-Up, and Drop-Off areas) they were characterized by personalization and flexibility, catering to specific user groups such as students, employees, seniors, and underserved populations. These services were described as highly technology-driven similar to the technology described from Waymo and Cruise--relying on mobile apps for booking and tracking, and advanced algorithms for route optimization. Public acceptance and regulatory challenges are significant hurdles that require strong advocacy, compliance efforts, and community engagement. Core examples included:

Campus Service:

- Scenario: Serving students, employees, and visitors within a campus.
- Stakeholders: National and regional authorities, local governments, and private entities.
- Challenges: Cost and funding, public acceptance, and technological limitations.

Low-Served Areas:

- Scenario: Providing services to low-served areas and populations.
- Stakeholders: Regional authorities and other governmental bodies.
- Challenges: Regulatory issues, stakeholder coordination, and operational challenges.

Suburban Areas:

- Deploying services in suburban areas, focusing on scalability and adaptability.
- Stakeholders: Regional and local authorities
- Challenges: Public trust, regulatory compliance, and operational efficiency.

Uber-like Services:

- Scenario: Implementing services similar to Uber, focusing on flexibility and user convenience.
- Stakeholders: Regional and local governments.
- Challenges: Public trust, regulatory issues, and operational management.

Senior Medical Transportation:

- Scenario: Offering transportation services to seniors requiring medical care.
- Stakeholders: Regional and local authorities, healthcare providers.
- Challenges: Regulatory compliance, public acceptance, and technological integration.

For Scheduled Services (Virtual stops, Fixed stops, Bus stops) the trend was focus on design for predictability and reliability, following fixed schedules. Experts related that they depended heavily on existing infrastructure like bus stops and dedicated lanes, and aim to optimize routes and schedules to enhance operational efficiency. These services were thought to require extensive stakeholder coordination, involving local and regional authorities and public transport operators.

Core examples included:

Urban Late-Night Routes:

- Scenario: Operating late-night fixed routes in urban areas.
- Stakeholders: Local/city government, public transport operators.
- Challenges: Public acceptance, infrastructure requirements, and cost management.

Airport-City Transit:

- Scenario: Providing mass transit between airports and city centers.
- Stakeholders: Regional authorities, public transport operators.
- Challenges: Stakeholder coordination, cost and funding, and market readiness.

Suburban Fixed Routes:

- Scenario: Connecting different areas within suburban regions through fixed routes.
- Stakeholders: Public transport operators, private entities.
- Challenges: Stakeholder coordination, cost and funding, public acceptance.

Public Service After Hours:

- Scenario: Offering services for general public use after regular public transport hours.
- Stakeholders: Regional and local authorities, public transport operators.
- Challenges: Stakeholder coordination, public trust, regulatory issues.

BRT Dedicated Lane Service:

- Scenario: Operating in Bus Rapid Transit (BRT) dedicated lanes, particularly at night and during weekends.
- Stakeholders: Local/city government, public transport operators.
- Challenges: Market readiness, cost and funding, infrastructure requirements.

4.1 Core Challenges and Stakeholder

In terms of the challenges faced by on-demand vs. scheduled services, as shown in Fig. 1, Cost and Funding is a major challenge for both entrepreneurial model groups, though less frequent compared in looking at scheduled services as opposed to on-demand service; conceivably this might be related to the perception that on-demand, door-to-door vehicles may have routes that are more complex, requiring more costly technology and service design. Regulatory and Legal Issues and Public Acceptance and Trust were also significant challenges for both, although they were seen as more significant in on-demand options given that they are usually managed by private vs. public operators.

How will the discussion continue from the breakout session? Who should know about the outcome and how will the information be disseminated? Who else might need to get involved in the discussion? Should the scope for the discussion be reconsidered? Which research projects should be considered or be reconsidered?

Core stakeholders that would need to be involved are shown in Fig. 2. For on-demand services, high involvement would be needed from government as well as the private sector but less support is perceived from academia and infrastructure stakeholders. For scheduled services, government and fleet operators are seen as more critical. National regulators are not seen as essential for any of these business models, and technology provider availability is not perceived as a challenge.

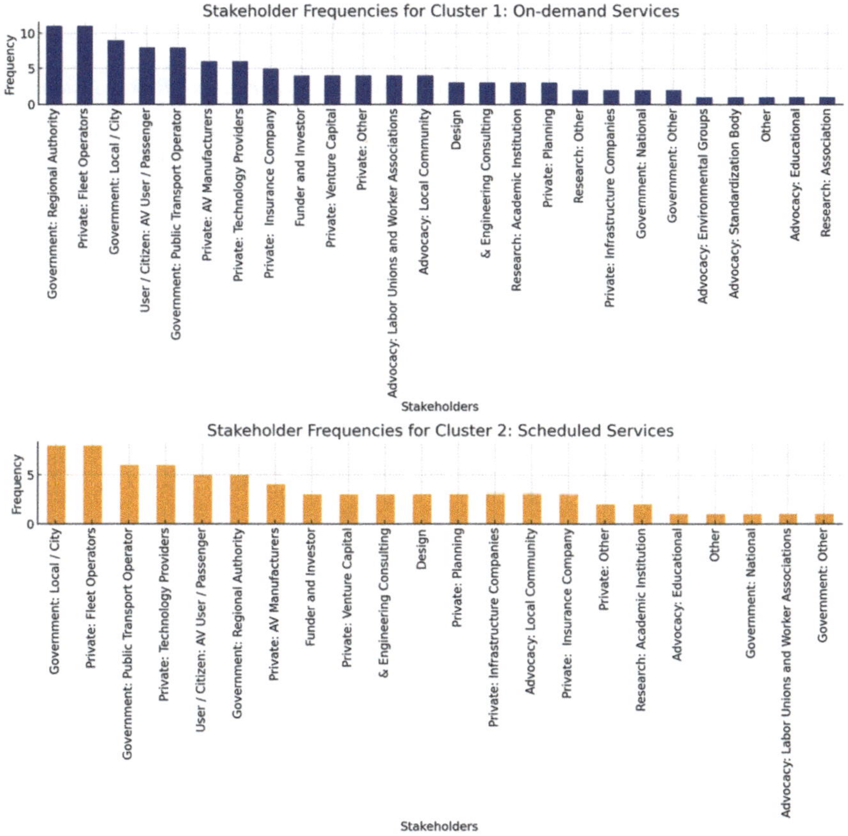

Fig. 2. Stakeholders Required by Entrepreneurial Cluster

4.2 Criteria Analysis

The matrices below in Table 2 and Table 3 assess each core example from the two clusters (On-demand and Scheduled services) based on six criteria: High Impact, Feasibility, Cost-Effectiveness, Timeliness, Risk, and Sustainability. Each criterion is scored on a scale from 1 to 5, with 1 being the lowest and 5 being the highest.

Table 2. Cluster 1: On-demand Services

Example	*High Impact*	*Feasibility*	*Cost-Effectiveness*	*Timeliness*	*Risk*	*Sustainability*
Campus Service	3	4	3	2	5	1
Low-Served Areas	4	2	5	1	3	5
Suburban Areas	4	4	4	3	3	3

(continued)

Table 2. (*continued*)

Example	*High Impact*	*Feasibility*	*Cost-Effectiveness*	*Timeliness*	*Risk*	*Sustainability*
Uber-like Services	5	5	1	5	4	2
Senior Medical Transportation	5	3	2	2	4	4

Table 3. Cluster 2: Scheduled Services

Example	*High Impact*	*Feasibility*	*Cost-Effectiveness*	*Timeliness*	*Risk*	*Sustainability*
Campus Service	3	4	4	1	3	5
Low-Served Areas	3	5	1	2	4	1
Suburban Areas	4	2	2	1	2	5
Uber-like Services	2	4	4	4	4	5
Senior Medical Transportation	3	1	4	2	4	2

For on-demand services, Senior Medical Transportation and Uber-like Services have the highest impact due to their wide applicability and user base, while Uber-like Services score high in feasibility due to established models and existing infrastructure. Low-Served Areas and Suburban Areas are seen as cost-effective due to potential for scaling. Uber-like Services have a high timeliness score, reflecting rapid deployment capability. Campus Service has high risk, likely due to technological and operational uncertainties. Low-Served Areas and Senior Medical Transportation score high in terms of sustainability, indicating a focus on long-term benefits.

For scheduled services, suburban Fixed Routes have the highest impact due to their potential to connect various regions efficiently. Airport-City Transit is highly feasible with existing infrastructure and stakeholder support. Urban Late-Night Routes and Public Service After Hours were cost-effective due to optimized scheduling. Public Service After Hours scores high, indicating quick implementation capability. Airport-City Transit has higher risk, possibly due to regulatory and logistical challenges. Suburban Fixed Routes and Public Service After Hours score high, focusing on long-term viability and environmental benefits.

5 Discussion

The findings from the expert surveys and the criteria matrix analysis illuminate the diverse opportunities and challenges associated with deploying autonomous vehicle (AV) technology. This section discusses key insights, emphasizing that there is no one-size-fits-all business model for AVs. Instead, a variety of business models must be adopted to ensure financial viability for companies, cities, and fleet management providers.

5.1 Variety of Business Models

The analysis indicates that both On-demand and Scheduled service models hold significant potential for AV deployment. On-demand models, such as Door-to-door and Pick-Up and Drop-Off areas (PUDOs), offer personalized and flexible services catering to specific user groups like students, employees, seniors, and underserved populations. These models leverage technology to optimize routes and enhance user experience, akin to the services provided by Waymo and Cruise. On the other hand, Scheduled services, including Virtual stops and Fixed stops, focus on predictability and reliability, often integrating with existing public transit infrastructure.

The diversity in AV business models underscores the necessity for tailored solutions based on local contexts, user needs, and regulatory environments. For instance, campuses might benefit from more scheduled services due to their confined geography and concentrated user base, while areas outside of an urban core, with less land-use diversity might find on-demand services more effective in connecting various regions efficiently.

5.2 Mutual Benefits for Public and Private Entities

A significant opportunity lies in the mutual benefits for public and private entities. Collaboration between these sectors can enhance public good and sustainability. Shared services can reduce private auto dependency, and data sharing between public and private sectors can optimize traffic management. For example, cities could "prime the pump" for automated services by providing mapping and 3D maps or sharing traffic and curb occupancy data, fostering an ecosystem in which AV companies can thrive.

Furthermore, dispelling the notion that AV operators benefit from increased traffic is crucial. Fleet operators and network efficiency managers have a vested interest in reducing vehicle miles/kilometers traveled (VMT/VKT). Efficient fleet management not only reduces operational costs but also aligns with sustainability goals by minimizing emissions and congestion. This cross-sector collaboration presents an excellent opportunity to streamline operations and enhance service delivery.

5.3 Targeted Use Cases for AV Services

Targeted use cases such as paratransit, disabled access, schools, campuses, and tourism locations present low-hanging fruit for AV business opportunities. For example, an expert highlighted the potential for AVs to transport individuals safely and efficiently in wine regions, allowing visitors to move between vineyards without the risk of driving under the influence of alcohol.

Such niche applications can demonstrate the feasibility and benefits of AV technology, fostering public trust and acceptance. By addressing specific needs and showcasing the advantages of AVs, these targeted use cases can pave the way for broader adoption and integration of AV services in diverse contexts.

5.4 Importance of Curb Design

Curb design is a critical consideration for the deployment of autonomous vehicles [38–42]. Well-designed curbs can significantly impact the safety, efficiency, and usability of AVs. Best practices include:

- Clear and Consistent Curb Markings: Using painted or raised markings to indicate crosswalks and loading zones helps AVs identify safe stopping and loading areas.
- Accessibility: Inclusive curb design with ramps, textured paving, and other features ensures that AVs can transport passengers and goods safely and effectively.
- Flexibility: Curbs should accommodate a range of vehicle types and sizes, including traditional cars, buses, and trucks, to prevent congestion and facilitate various transportation needs.
- Safety: Prioritizing safety for all road users by incorporating features like separated bike lanes and protected pedestrian crossings reduces the risk of accidents and injuries.

This strategy is consistent with International Association of Transportation Regulator's (IATR) "Best Practices, Guiding Principles & Model Regulations for Robotaxis" emphases that self-driving connected and autonomous electric vehicles (S-CAEVs) must exceed current safety standards for taxis and for-hire vehicles [43, 44]. The best practices call for transitioning to zero-emission fleets, mandating accessibility features for passengers with disabilities, and instituting robust data privacy protections. The principles stress equitable access to AV services, prioritizing vulnerable communities and advocating for business models that accommodate low-income and unbanked populations.

Integrating Vision Zero principles into AV implementation is essential. Vision Zero is a global initiative to eliminate all traffic fatalities and severe injuries by designing a transportation system that prioritizes safety, equity, and human error mitigation. As AVs become more prominent, prioritizing safety in their design and deployment is critical. While stakeholders may focus on automated enforcement or alternative modes of transportation, AVs should also be central to Vision Zero strategies, reducing traffic fatalities and injuries. Incorporating Vision Zero principles ensures AVs are deployed to reduce traffic fatalities and injuries, creating safer roads for all.

5.5 Civic Investment and Infrastructure

There is a clear need for civic investment in supporting infrastructure to facilitate the deployment of automated shared services. Cities can accelerate AV adoption by investing in reliable charging infrastructure, dynamic curb use, AV dedicated lanes, and sharing digital information about construction sites and emergencies. Providing these elements can significantly enhance the operational efficiency of AV services and foster a conducive environment for private sector investment and innovation.

6 Conclusion

By applying a structured approach to business model evaluation, this study identifies both the opportunities and challenges in deploying autonomous vehicle (AV) technology. The analysis reveals the necessity for diverse business models tailored to specific contexts, as no single model can address the varied needs and circumstances of different regions and user groups. Both On-demand and Scheduled service models, including Door-to-door services, Pick-Up and Drop-Off areas (PUDOs), Virtual stops, and Fixed stops, present viable frameworks for integrating AVs into urban environments.

The mutual benefits for public and private entities are significant, as collaboration can enhance public good and sustainability. Shared services can reduce the number of cars on the road and optimize traffic management—and potentially lead to lower auto dependency and less need for car ownership. Fleet operators and AV network managers, both public and private, can aim to reduce vehicle miles/kilometers travelled (VMT/VKT), aligning operational efficiency and network performance with sustainability goals and reduced environmental impact per VMT/VKT.

Targeted use cases such as paratransit, disabled access, schools, campuses, and tourism locations demonstrate the practical applications of AV technology. These niche opportunities can foster public trust and acceptance, showcasing the advantages of AVs in improving safety, efficiency, and service delivery. Curb design also emerges as a crucial factor in the deployment of AVs, impacting safety, efficiency, and usability. Best practices in curb design, such as clear markings, accessibility features, flexibility to accommodate various vehicle types, and safety measures for all road users, are essential for successful AV integration.

Ultimately, civic investment in supporting infrastructure is vital for the successful deployment of automated shared services. Cities can facilitate AV adoption by investing in reliable charging infrastructure, dynamic curb use, AV dedicated lanes, and sharing digital information about construction sites and emergencies. These investments can significantly enhance operational efficiency and create a conducive environment for private sector innovation.

In conclusion, the structured evaluation approach underscores the transformative potential of AV technology. By addressing the identified challenges and leveraging entrepreneurial innovation, AVs can pave the way for more resilient and sustainable transportation systems, ultimately contributing to improved urban mobility.

References

1. Riggs, W., Pande, A.: On-demand microtransit and paratransit service using autonomous vehicles: Gaps and opportunities in accessibility policy. Transp. Policy **127**, 171–178 (2022). https://doi.org/10.1016/j.tranpol.2022.07.024
2. Riggs, W., Beiker, S.A.: Business models for shared and autonomous mobility. In: Road Vehicle Automation 7, Meyer, G., Beiker, S. (eds.), in Lecture Notes in Mobility. Cham: Springer International Publishing, pp. 33–48 (2020). https://doi.org/10.1007/978-3-030-52840-9_4
3. Appleyard, B., Riggs, W.: A Conceptual Framework for Livability, Sustainability and Equity in the Era of Automated Transit. SSRN. Accessed 10 Aug 2022. https://papers.ssrn.com/sol3/papers.cfm?abstract_id=3813920

4. Mahmoodi Nesheli, M., Li, L., Palm, M., Shalaby, A.: Driverless shuttle pilots: Lessons for automated transit technology deployment. Case Studies on Transport Policy **9**(2), 723–742 (2021). https://doi.org/10.1016/j.cstp.2021.03.010
5. Alessandrini, A., Campagna, A., Delle Site, P., Filippi, F., Persia, L.: Automated vehicles and the rethinking of mobility and cities. Transportation Research Procedia **5**, pp. 145–160 (2015)
6. Riggs, W.: Disruptive transport: driverless cars, transport innovation and the sustainable city of tomorrow. in Routledge Equity, Justice and the Sustainable City Series. Routledge, London (2019)
7. Prince, S., Chapman, S., Cassey, P.: The definition of entrepreneurship: is it less complex than we think? Int. J. Entrep. Behav. Res. **27**(9), 26–47 (2021). https://doi.org/10.1108/IJEBR-11-2019-0634
8. Parrish, B.D.: Sustainability-driven entrepreneurship: principles of organization design. J. Bus. Ventur. **25**(5), 510–523 (2010). https://doi.org/10.1016/j.jbusvent.2009.05.005
9. Carlton, I.: Histories of Transit-Oriented Development: Perspectives on the Development of the TOD Concept (2009). Accessed 04 Feb 2025. https://escholarship.org/uc/item/7wm9t8r6
10. UITP. The Rural Mobility Challenge for Public Transport: How Combined Mobility Can Help. UITP (2022). https://cms.uitp.org/wp/wp-content/uploads/2022/02/Knowledge-Brief-Rural-Mobility_FEB2022-web.pdf
11. Cornet, H., Riggs, W., Gkemou, M., Dreher, S.: On international collaboration within research projects for large-scale piloting of shared automated mobility. In: Shared Mobility Revolution: Pioneering Autonomous Horizons, Springer Nature Switzerland Cham, pp. 233–246 (2024). Accessed 13 Jan 2025. https://library.oapen.org/bitstream/handle/20.500.12657/96080/9783031717932.pdf?sequence=1#page=230
12. Alliance for Automotive Innovation. Ready to Launch Autonomous Vehicles in the U.S. Tracking the current (and future) AV landscape. Alliance for Automotive Innovation, Washington D.C. (2022). https://www.autosinnovate.org/posts/papers-reports/AV%20Report.pdf
13. Bellan, R.: Cruise can charge for driverless robotaxi rides in SF. TechCrunch. Accessed 18 Apr 2024. https://techcrunch.com/2022/06/02/cruise-can-finally-charge-for-driverless-robotaxi-rides-in-san-francisco/
14. Dow, J.: GM Cruise takes first fares for paid driverless taxi in San Francisco. Electrek. Accessed 01 Mar 2023. https://electrek.co/2022/06/23/gm-cruise-takes-first-fares-for-paid-driverless-taxi-in-san-francisco/
15. Ewing, J., Tan, E.: G.M. Will Stop Developing Self-Driving Taxis. The New York Times (2024). Accessed 04 Feb 2025. https://www.nytimes.com/2024/12/10/business/gm-cruise-robotaxi.html
16. Roth, E.: Waymo can now Provide Fully Driverless Rides in San Francisco. The Verge. Accessed 18 Apr 2024. https://www.theverge.com/2022/11/19/23467784/waymo-provide-fully-driverless-rides-san-francisco-california
17. Hawkins. Waymo Eyes Traffic-Choked Los Angeles for its Next Robotaxi Service - The Verge. The Verge. Accessed 01 Aug 2024. https://www.theverge.com/2022/10/19/23410677/waymo-los-angeles-autonomous-robotaxi-service-launch
18. Bellan, R.: Waymo Reportedly Testing Robotaxis in 10 New Cities in 2025. TechCrunch. Accessed 04 Feb 2025. https://techcrunch.com/2025/01/29/waymo-reportedly-testing-robotaxis-in-10-new-cities-in-2025/
19. Tan, E.: Waymo's Robot Taxis Are Almost Mainstream. Can They Now Turn a Profit?. The New York Times (2024). Accessed 04 Feb 2025. https://www.nytimes.com/2024/09/04/technology/waymo-expansion-alphabet.html
20. Korosec. Nuro's Self-Driving Vehicles to Deliver Prescriptions for CVS Pharmacy. TechCrunch. Accessed 29 Jun 2020. https://social.techcrunch.com/2020/05/28/nuros-self-driving-vehicles-to-delivery-prescriptions-for-cvs-pharmacy/

21. Autobody News. Waymo Dominates Robotaxi Market, Faces Growing Competition. Autobody News. Accessed 04 Feb 2025. https://www.autobodynews.com/news/waymo-dominates-robotaxi-market-faces-growing-competition
22. Nolan, S.: Top 10: Autonomous Vehicles. Accessed 04 Feb 2025. https://evmagazine.com/top10/top-10-autonomous-vehicles
23. Olsen. Quiet, clean and smart: New electric autonomous bus is ready for riders. MSUToday | Michigan State University. Accessed 18 Apr 2024. https://msutoday.msu.edu/news/2022/new-electric-autonomous-bus-accepts-riders
24. Zukowski, D.: Autonomous Shuttles will Replace Monorail System in Jacksonville, Florida. Smart Cities Dive (2022). Accessed 18 Apr 2024. https://www.smartcitiesdive.com/news/jacksonville-florida-autonomous-shuttle-monorail-people-mover-transit-oriented-development/629612/
25. Welle, E.: May Mobility Launches Free Autonomous Vehicle Shuttle Service in Ann Arbor. The Michigan Daily. Accessed 18 Apr 2024. http://www.michigandaily.com/news/autonomous-vehicle-start-up-may-mobility-launches-free-passenger-shuttle-service-at-ann-arbor-farmers-market/
26. Leahy. New Treasure Island Robo-Bus Trial Secures Florida-Based Operator. The San Francisco Standard. Accessed 04 Feb 2025. https://sfstandard.com/2022/10/25/treasure-island-robo-bus-trial-secures-florida-based-operator-beep/
27. Leahy. We Rode San Francisco's First Driverless Bus. It Was Jarring. The San Francisco Standard. Accessed 04 Feb 2025. https://sfstandard.com/2023/08/17/we-rode-san-franciscos-first-driverless-bus-it-was-jarring-and-not-exactly-driverless/
28. Alkim. Seeing is believing: Riding with Cruise and Waymo in San Francisco - Connected Automated Driving. Connected and Autonomous Driving EU. Accessed 04 Feb 2025. https://www.connectedautomateddriving.eu/blog/seeing-is-believing-riding-with-cruise-and-waymo-in-san-francisco/
29. Cornet, H., et al.: Report on international cooperation opportunities. FAME Deliverable D3.2 (2024)
30. Juliussen, E.: EE Times Europe - Autonomous Vehicles: How Is Europe Doing?. EE Times. Accessed 01 Aug 2024. https://www.eetimes.eu/autonomous-vehicles-how-is-europe-doing/
31. Ride the Future. Ride the future," Ride the Future. Accessed 04 Feb 2025. https://ridethefuture.se/in-english/
32. Swiss Transit Lab. Furttal Project. Project description. Accessed 04 Feb 2025. https://www.swisstransitlab.ch/de/projekte/furttal/
33. National Police Agency, Japan. Automated Driving|National Police Agency," National Police Agency. Accessed 04 Feb 2025. http://www.npa.go.jp/english/bureau/traffic/selfdriving.html
34. Ministry of Economy, Trade and Industry, Japan. Operation of Automated Driving Vehicle with a Level 4 System Approved for the First Time in Japan. Accessed 04 Feb 2025. https://www.meti.go.jp/english/press/2023/0512_002.html
35. Softbank News. Level 4 Autonomous Bus Service to Start Near Haneda Airport. SoftBank News. Bringing IT closer to home. Accessed 04 Feb 2025. https://www.softbank.jp/en/sbnews/entry/20240627_01
36. Xinhua News. Xinhua Headlines: China's autonomous driving enters 'fast lane' with commercial operations-Xinhua. Xinhua Headlines. Accessed 01 Aug 2024. https://english.news.cn/20220815/12f18d011f5247a18cca5dc468905c37/c.html
37. Butcher, L.: Robosweeper ready to take to the streets of Guangzhou. ADAS & Autonomous Vehicle International. Accessed 01 Aug 2024. https://www.autonomousvehicleinternational.com/news/testing/robosweeper-ready-to-take-to-the-streets-of-guangzhou.html
38. Riggs, W., Appleyard, B., Johnson, M.: A design framework for livable streets in the era of autonomous vehicles. Urban, Planning and Transport Research **8**(1), 125–137 (2020). https://doi.org/10.1080/21650020.2020.1749123

39. Riggs, W., Appleyard, B., Johnson, M.: A design framework for livable streets in the era of autonomous vehicles. In: Proceedings of the 98th Annual Meeting Transportation Research Board, Washington, D.C. (2019)
40. Riggs, W., Ruhl, M., Rodier, C., Baumgardner, W.: Designing streets for autonomous vehicles. In: Road Vehicle Automation 6, Meyer, G., Beiker, S. (eds.), in Lecture Notes in Mobility. , Springer International Publishing, pp. 111–122 (2019)
41. Schlossberg, M., Riggs, W.W., Millard-Ball, A., Shay, E.: Rethinking the Street in an Era of Driverless Cars. UrbanismNext (2018). https://urbanismnext.uoregon.edu/files/2018/01/Rethinking_Streets_AVs_012618-27hcyr6.pdf
42. Appleyard, B., Riggs, W.: Designing for street livability in the era of driverless cars. Transportation Research Interdisciplinary Perspectives **21**, 100868 (2023). https://doi.org/10.1016/j.trip.2023.100868
43. Daus, M.: The Robotaxi Movement is in Full Gear: 'ARTS-fully Done TRB!." Black Car News. Accessed 04 Feb 2025. https://www.blackcarnews.com/article/the-robotaxi-movement-is-in-full-gear-arts-fully-done-trb
44. IATR. Best Practices, Guiding Principles & Model Regulations for Robotaxis. International Association of Transportation Regulator. Accessed 04 Feb 2025. https://www.dropbox.com/scl/fi/6ocmid3vwezj3g6490277/IATR-s-Best-Practices-Guiding-Principles-Model-Regulations-for-Robotaxis.pdf?dl=0&rlkey=ydz4k4tam2aehg82ogdcez5kw&st=3gcyvizm

Shaping Future Mobility: Mapping the European CCAM Stakeholder Ecosystem

Carolin Zachäus(✉) and Alexander Egoldt

VDI/VDE Innovation + Technik GmbH, Steinplatz 1, 10623 Berlin, Germany
{carolin.zachaeus,alexander.egoldt}@vdivde-it.de

Abstract. This chapter presents a detailed analysis of the European stakeholder ecosystem for Connected, Cooperative, and Automated Mobility (CCAM), drawing on findings from the FAME project. Using stakeholder mapping and network science methodologies, it uncovers structural features, strategic gaps, and key brokers essential for advancing large-scale CCAM deployment in Europe. The chapter examines stakeholder types, geographic imbalances, thematic clusters, and coordination mechanisms, with particular attention to underrepresented groups such as public transport operators, MaaS providers, and civil society actors, as well as stakeholders from Eastern Europe. It highlights actionable strategies—empowering strategic connectors, activating associations and city networks, aligning national and EU-level initiatives, and bridging research with deployment—to build a more inclusive, integrated, and deployment-ready CCAM ecosystem, setting the stage for future initiatives.

Keywords: Connected and Automated Mobility · Autonomous Driving · CCAM Partnership · CCAM Ecosystem · Innovation Policy · Research and Innovation Activities

1 Introduction

Connected, Cooperative and Automated Mobility (CCAM) is central to Europe's vision for safe, efficient, and sustainable transport. By integrating digital technologies, automation, and coordinated stakeholder engagement, CCAM has the potential to transform the mobility system. Yet, large-scale deployment requires more than technological excellence—it depends on a well-connected stakeholder ecosystem that spans public authorities, industry, research, implementers, infrastructure provider, transport operators and civil society.

European Initiatives and Current Developments

CCAM is at the heart of Europe's transport innovation agenda, bringing together diverse technologies, stakeholders, and governance structures to deliver safer, sustainable, and efficient mobility solutions. Key initiatives—such as the Horizon Europe CCAM Partnership—are reinforced by new alliances, industrial strategies, and policy measures designed to accelerate the shift towards connected and automated mobility. Hereby,

G. Meyer and S. Beiker (Eds.): ARTSymposium 2024, LNMOB, pp. 63–78, 2026.
https://doi.org/10.1007/978-3-032-12374-9_5

targeted action plans for the automotive sector, proposals for integrated European partnerships, and collaborative frameworks for connected and automated vehicles define the policy, research, and industrial environment for Europe's mobility transition.

CCAM Partnership

Launched on 23 June 2021 under Horizon Europe, the CCAM Partnership unites industry, research bodies, public authorities, and user associations to speed the development and deployment of connected and automated mobility. Its goal is a user-centred, inclusive mobility system that enhances safety, reduces congestion, and lowers environmental impact. The partnership supports collaborative research, cross-border demonstrations, and alignment of research and innovation (R&I) with policy goals, public trust and market readiness.[1]

Work is organised into seven thematic clusters:

- Large-scale demonstrations – Testing automated mobility in real-world contexts.
- Vehicle technologies – Advancing hardware and software for connected and automated vehicles.
- Validation – Ensuring safety, reliability, and regulatory compliance.
- Integration – Achieving interoperability between vehicles, infrastructure, and services.
- Key enabling technologies – Focusing on data, AI, and cybersecurity.
- Societal aspects – Addressing public acceptance, inclusivity, and behavioural impacts.
- Coordination – Aligning strategies and stakeholder engagement, and supporting knowledge exchange.

Established under the CCAM Partnership, the CCAM States Representatives Group (SRG)[2] advises on partnership objectives and ensures alignment with national policies and programmes. It strengthens coherence across policies and programmes on European and national level, facilitates the sharing of best practices and supports deployment through workshops and dissemination events.

Industrial Action Plan for the European Automotive Sector

Published in March 2025, the plan emphasizes global pressures on electrification, automation, and competitiveness in the automotive sector. It defines five priorities, namely Innovation and digitalisation, Clean mobility, Competitiveness and supply chain resilience, Skills and social dimension, Boosting market access, ensuring a level playing field and guaranteeing economic security. Measures range from workforce reskilling R&I in autonomous vehicles and batteries, with a strong focus on leadership in AI-enabled, software-defined mobility and integration with EU space-based services. As part of it the specifically mentioned European Connected and Autonomous Vehicle Alliance will develop shared software and hardware foundations for connected and automated vehicles, including an open platform for Software-Defined Vehicles (SDV), in-vehicle computing architecture for SDV and AI solutions for the automotive industry.

---

[1] https://www.ccam.eu/

[2] https://www.ccam.eu/what-is-ccam/governance/ccam-states-representatives-group/

Further key activities include joint R&I projects, shared testbeds, pilot deployments, and interoperability standards.[3]

European Forum on Automated Transport
Established in 2024, the European Forum on Automated Transport (EFAT) facilitates structured exchanges among EU and neighbouring states and anchors the Joint Commitments of the Member States. These commitments include: establishing a joint agenda of policy and regulatory issues; sharing information on national strategies, roadmaps, regulations, and programmes; exchanging best practices on integration into national frameworks; fostering mutual understanding of national implications once automated vehicles are deployed (traffic enforcement, roadworthiness, infrastructure, licensing, public acceptance).[4]

European Coalition of Local and Regional Authorities for Automated Road Transport Services
Launched in June 2025, this coalition mobilises local and regional authorities to integrate automated transport into public transit systems. The inaugural event on 27 June in Paris—organised by the French government, France Véhicules Autonomes (FVA), and Norway's Ruter, with support from POLIS, EMTA, EIT Urban Mobility, Eurocities, UITP, PAVE Europe, ERTICO, and the CCAM Partnership—marked the first step in building a European coalition of transport providers for automated mobility. Its goal is to create a critical mass of demand for automated services and ensure that European cities and public transport operators (PTO) lead in shaping mobility's future. It applies a user-centric approach, integrating automation into existing public transport networks to improve service quality, reduce costs, and support sustainable mobility planning.

Although Europe has a strong R&I base in automotive and transport technologies, the CCAM ecosystem remains fragmented, with gaps in stakeholder inclusion and regional representation. The FAME project addressed these challenges by conducting a comprehensive stakeholder mapping and network analysis to map Europe's CCAM ecosystem, identify core actors, and highlight opportunities to strengthen cohesion and inclusivity. The analysis examines collaboration patterns, the distribution of roles across sectors, and the geographic spread of stakeholders, identifying both highly connected "hubs" and underrepresented groups that may require targeted engagement. It assesses the density and diversity of relationships, the presence of cross-border links, and the role of intermediary organisations in bridging industry, research, and governance. Findings provide actionable insights for enhancing stakeholder participation, aligning national and European R&I priorities, and accelerating the transition from research outcomes to market-ready solutions. By revealing where collaboration is strong and where gaps persist—whether in technical interoperability, policy alignment, or societal engagement—the analysis supports strategic decision-making for both public and private actors.

[3] European Commission, COM(2025) 95 final, Industrial Action Plan for the European automotive sector, 05.03.2025, https://transport.ec.europa.eu/document/download/89b3143e-09b6-4ae6-a826-932b90ed0816_en?filename=Communication%20-%20Action%20Plan.pdf.

[4] https://www.connectedautomateddriving.eu/wp-content/uploads/2025/05/EUCAD25_EFAT.pdf.

2 Methodological Framework

The methodology for mapping the CCAM ecosystem follows a comprehensive, exploratory, and data-driven approach that integrates both quantitative and qualitative analysis that has been performed under der Horizon Europe project[5]. It is structured around the collection, cleaning, classification, and interpretation of a dataset on stakeholders, projects, and associations involved in CCAM-related research and strategic coordination activities across Europe.

Data Acquisition and Preparation

The core dataset includes structured information from the CCAM Knowledge Base[6], CORDIS database[7], and the membership directories of European-level associations, platforms, and clusters. Stakeholders from EU-funded R&I programs FP7, Horizon 2020, and Horizon Europe were included. Each actor was geolocated and classified into one of 15 stakeholder categories defined by the CCAM Partnership. Associations were given a "dual role"—as strategic connectors and independent actors—allowing for a nuanced depiction of their function within both research and strategic dimensions.

Network Analysis and Visualisation

Network analysis offers a powerful lens through which to explore and understand the complex CCAM ecosystem, characterized by a wide variety of stakeholders operating across different levels. This analysis adopts a multi-level approach, addressing both research-related and strategic dimensions to provide a comprehensive view of stakeholder relationships.

At the research level, network analysis helps uncover relationships through the shared participation of stakeholders in collaborative research projects. In contrast, the strategic level focuses on connections formed within associations, formal partnerships, and informal interest groups—such as the CCAM Partnership.

To differentiate the intensity of these connections, a weighting system was applied using a scale from 1 to 5. Collaborations at the research level are assigned a weight of 5, acknowledging their intensive nature. However, these ties tend to be dependent on active engagement with specific topics; once a research theme loses relevance, the associated collaborations often dissolve. Conversely, connections at the strategic level are generally broader and less binding, and are thus assigned a weight of 1. Although strategic networks can offer vast reach, they often include competitors and do not inherently promote deep or reciprocal relationships—particularly within large associations. Nonetheless, their significance lies in enabling extensive networking opportunities and amplifying the collective voice of the ecosystem. An intermediate weighting of 2 is given to connections within the CCAM Partnership. These structures are seen as more effective facilitators of cooperation, primarily due to frequent networking activities and dedicated measures to foster collaboration specially focusing in CCAM. In addition, these networks provide avenues for stakeholders to contribute to shaping the European Commission's CCAM R&I strategy, further enhancing their strategic value.

[5] FAME Deliverable D3.5: Report on Impact and Outcomes of Stakeholder Engagement.

[6] https://www.connectedautomateddriving.eu/projects/findproject/

[7] https://cordis.europa.eu/

The network analysis also involved a classification of research-based connections. This classification was guided by the seven thematic clusters of the CCAM Partnership: large-scale demonstration, vehicle technologies, validation, integrating CCAM in the transport system, key enabling technologies, societal aspects and people needs, and coordination. Stakeholders were categorized according to their type, following the framework used by the CCAM Partnership. These include research institutions, universities, automotive manufacturers and suppliers, stakeholder associations at the European level, national and regional platforms and clusters, various levels of transport authorities, ministries, public transport operators, and providers of physical and digital infrastructure or related services.

To evaluate the relevance and positioning of stakeholders within the network, several analytical parameters were employed[8]. These included degree of centrality, closeness centrality, and betweenness centrality (for definition see Table 1).

Table 1. Definition of analytical network parameters; degree of centrality, closeness centrality, betweenness centrality.

Degree of centrality
Degree of centrality defines the total number of edges an actor has. It is a simple measure of the strength of links between an actor and all the other actors in the ecosystem. A high value of degrees suggest that an actor is well connected and thus has access to more information.
Closeness centrality
Closeness denotes the spatial distance of an actor to all other actors, indicating which actors are the most influential at the "centre" of the network. It furthermore serves as an index of the viability of access to information possessed by the actor since a higher distance amongst actors implies weaker links and thus reduced information flows. This means, an actor with high closeness is less reliant on mediation by other actors to reach even further actors.
Betweeness centrality
Betweenness provides a measure of the structural dependence of a pair of actors on a third actor, showing how often an actor is on the shortest path to other actors. Thus, when the betweenness is high, the actor has access to different and new flows of information through partners who have participated in mutually unconnected projects.

While the degree of centrality is a local measure based on immediate connections, closeness and betweenness are considered global measures. Closeness centrality helps determine how independently a stakeholder can access the broader network, while betweenness highlights those who link different parts of the ecosystem. Furthermore, eigenvector centrality was considered to assess how close an actor is to the network's core—namely, the most active and influential participants.

[8] Stegbauer, C. (Hg.) (2010). Netzwerkforschung: Vol. 2. Netzwerkanalyse und Netzwerktheorie: Ein neues Paradigma in den Sozialwissenschaften (2. Aufl.). Wiesbaden: VS Verlag für Sozialwissenschaften.

The technical execution of the network analysis and visual representation was carried out using Gephi[9], a dedicated network analysis software. Complementary visualizations, such as the distribution of stakeholder types and the number of active stakeholders per country, were created using Tableau and Microsoft Excel.

Limitations

While network analysis offers valuable insights into stakeholder dynamics, several limitations warrant consideration. First, the analysis incorporates connections that are potentially reactivatable—relationships formed through past collaborations that could be renewed. However, many of these ties exist at the individual rather than institutional level, making them inherently less stable. As a result, the network should be viewed as a "space of possibilities" rather than a definitive map of ongoing activity.

The scope of the analysis is constrained by data availability and quality: difficulties in identifying collaborations at national or international levels have led to a focus on European-level, publicly available datasets—particularly EU-funded R&I projects.

Automated classification methods, while essential for processing large datasets, carry a risk of occasional misclassification—both in assigning projects to thematic clusters and in categorizing stakeholder types. Moreover, frequent membership changes in associations such as the CCAM Partnership and CAD Network mean the analysis reflects only a snapshot in time, though the overarching trends are expected to remain valid.

3 Mapping the Evolving European CCAM Ecosystem Through Network Analysis

The European CCAM ecosystem is a complex and rapidly evolving network of stakeholders spanning multiple sectors. A comprehensive network analysis reveals a landscape centered around collaborative research while increasingly incorporating actors from adjacent domains such as telecommunications, data and AI, user advocacy, and Mobility-as-a-Service (MaaS).

At the heart of this ecosystem lie European R&I projects and strategic partnerships and associations, serving as the primary hubs of connectivity across sectors. Around these central structures, a broader network of approximately 4,000 stakeholders takes shape, reflecting the diversity and scale of the CCAM landscape (Fig. 1).

A key finding of the analysis is the emergence of dense, internally connected clusters formed by large associations—such as UITP (public transport), IRU (road transport unions), ERF (road infrastructure operators), and ECG (vehicle logistics). While these groups show strong internal cohesion, many of their members remain loosely connected to the broader ecosystem. These "association clouds" represent significant latent potential. With targeted initiatives—such as matchmaking events, cross-sector task forces, or co-funded pilot projects—these under-connected clusters could be more effectively integrated into the European R&I landscape.

In contrast, actors directly involved in European R&I projects and the CCAM partnership exhibit a higher degree of cross-sector integration. Their members are embedded

[9] https://gephi.org/

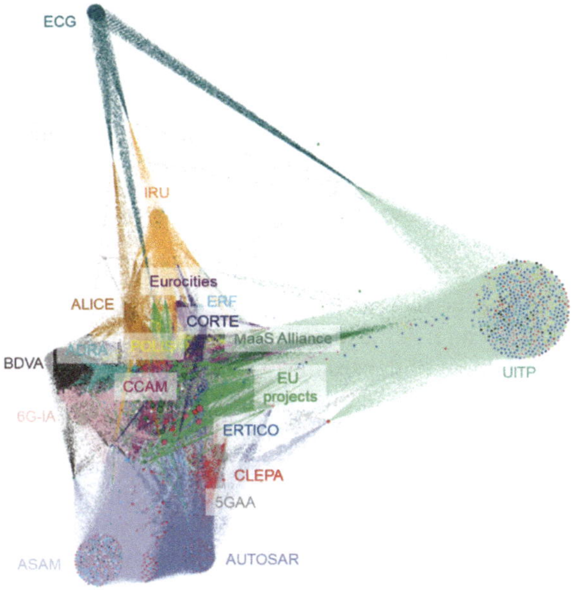

Fig. 1. Overview of entire CCAM ecosystem, addressing all interaction levels

in the ecosystem's core and actively engage across multiple domains. Notably, recent efforts to include vertical-domain associations have begun to create new bridges: 5GAA and 6G-IA link telecom providers with mobility stakeholders; BDVA and ADRA bring in AI and data innovation actors; and the MaaS Alliance connects service integrators with traditional transport providers. These connections form fine connective threads across previously siloed areas, enhancing the overall coherence of the ecosystem.

The analysis also highlights several key players based on structural network metrics. Organizations such as VTT, TNO, and ERTICO show the highest weighted degree, indicating broad participation in projects and associations. Siemens, TNO, and ICCS rank highest in closeness centrality, making them strategically positioned to quickly interact with any part of the network. Meanwhile, CERTH, Siemens, and ERTICO lead in betweenness centrality, serving as bridges between otherwise disconnected subcommunities. Importantly, public-sector entities and cross-sector organizations—like the City of Lisbon and UITP—emerge as influential connectors, showing that cities and associations play a dual role as contributors to innovation and ecosystem enablers.

Yet, integration challenges persist. Some actors remain highly embedded within their own association clusters but have few or no connections beyond them, limiting their influence across the wider CCAM ecosystem. For example, many members of UITP maintain strong internal links but lack outward-facing partnerships, underscoring the need for outreach strategies that extend beyond institutional boundaries. Additionally, a persistent underrepresentation of stakeholders from Eastern Europe highlights the need for greater engagement with regional infrastructure players, telecom providers, and innovation clusters to ensure a more geographically balanced and resilient network.

Overall, this analysis confirms that research-driven collaboration remains the backbone of the European CCAM ecosystem. At the same time, the increasing inclusion of actors from infrastructure, data, telecommunications, and service sectors is building the foundations for a more integrated, inclusive, and deployment-ready mobility landscape across Europe.

4 Understanding the Structure of the European CCAM Research Ecosystem

An in-depth analysis of over 160 EU-funded projects—spanning from FP7 to Horizon Europe—reveals the structure, key actors, and thematic focus areas within the European (CCAM) research landscape. With participation from more than 1,300 stakeholders, the network reflects both thematic clustering and a broad diversity of actors.

Projects integrating vehicles into the transport system emerged as the most central and densely connected cluster. These initiatives frequently unite cities, vehicle manufacturers, telecom providers, and system integrators, establishing them as primary drivers of ecosystem-wide connectivity. Clusters focused on vehicle and key enabling technologies are also well-represented, sustaining the technical backbone of the network. However, areas such as societal aspects, user acceptance, and validation remain underrepresented—highlighting a gap in the broader effort to foster user trust and regulatory readiness. This indicates that these topics—though crucial for long-term deployment and public trust—are still not prioritised within the broader research community.

The stakeholder landscape of EU CCAM R&I projects is predominantly research-driven. Research institutes constitute the largest share of participants, representing approximately 29% of all project partners. Universities follow closely, making up around 16% of the involved actors. While companies from the automotive supply chain contribute a modest 10%, their presence signals steady, though still secondary, industry engagement. Infrastructure stakeholders—those responsible for developing and operating the physical and digital backbone of CCAM—also play a growing role, accounting for 10% of project participants and indicating increasing integration of deployment-oriented actors into the research domain.

Specifically, organisations such as ERTICO, TNO, and VTT emerge as structurally central actors within the research network. Their high levels of closeness and betweenness centrality indicate a strong ability to connect different parts of the ecosystem. These actors play a pivotal role in facilitating knowledge flow, bridging thematic clusters, and enabling cross-sector synergies—positioning them as key drivers of system integration, standardisation, and the translation of research into applied outcomes.

At the same time, the research-level network shows a noticeable underrepresentation of critical stakeholder groups, including mobility providers, national and local authorities, public transport operators, and civil society organisations. Their limited presence suggests that the perspectives of end users and deployers are not yet fully embedded in the research agenda. This disconnect risks narrowing the relevance and public acceptance of CCAM solutions.

Coordination activities, although involving relatively few projects, occupy central positions in the network. Their structural importance highlights the need for overarching governance mechanisms and shared frameworks that maintain coherence across the fragmented and expanding CCAM landscape. These coordination projects serve as essential connectors, supporting cross-project alignment and promoting open knowledge exchange.

Furthermore, while automotive suppliers and OEMs maintain strong technical contributions to research, their involvement in broader ecosystem-level discussions remains limited. Their focus has largely remained on domain-specific innovation, with comparatively lower engagement in areas such as societal acceptance, user needs, and system-level integration. This points to an imbalance in the current research landscape and underscores the need for more holistic collaboration across disciplines and sectors.

Developments in the CCAM Research Landscape Under Horizon Europe
The Horizon Europe (HE-FP9) framework marks a continued expansion and strengthening of the CCAM research ecosystem. With 37 funded projects, the network has welcomed new actors while reinforcing the roles of established stakeholders. Large-scale demonstration projects dominate the collaboration landscape, accounting for about 30% of activities, closely followed by vehicle integration projects at 27%. Other thematic areas—including validation (13%), key enabling technologies (9%), and vehicle technologies (8%)—also hold significant shares, whereas societal aspects, user acceptance (7.5%), and coordination activities (6%) remain less prominent (Fig. 2).

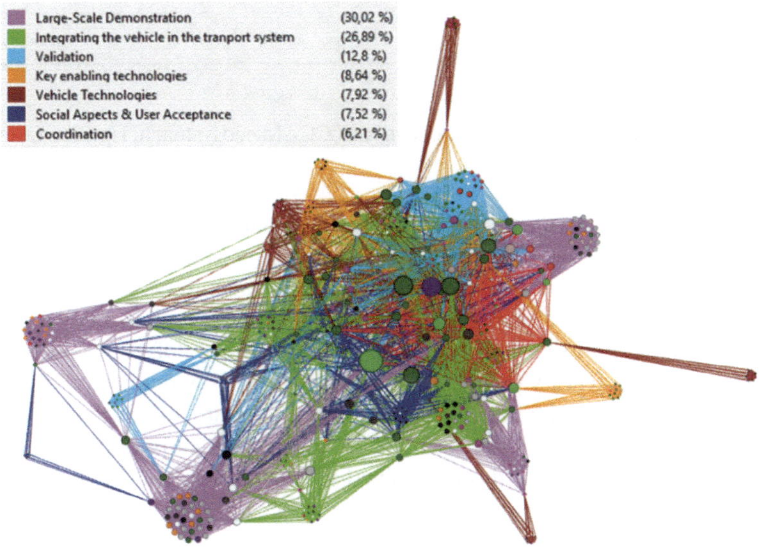

Fig. 2. Latest CCAM ecosystem development on European R&I level in Horizon Europe (FP9)

Structural analysis highlights the central role of research organisations and pan-European associations (Fig. 3, Table 2). ERTICO again emerges as the key connector, leading across all centrality metrics and bridging industry, academia, and the public sector. Institutions such as ICCS, CERTH, TNO, and Vicomtech follow closely, linking multiple clusters and supporting both technical and strategic dimensions. IDIADA, in particular, has gained strong prominence in the updated landscape, playing a pivotal role in projects focused on testing, validation, and certification—critical areas for scaling CCAM deployment. Public research bodies and universities—including TU Eindhoven, VTT, DLR, and VTI—remain essential anchors for specialised consortia and thematic collaboration. Newcomers like PAVE Europe and EMT Madrid have also established themselves as cross-sector enablers.

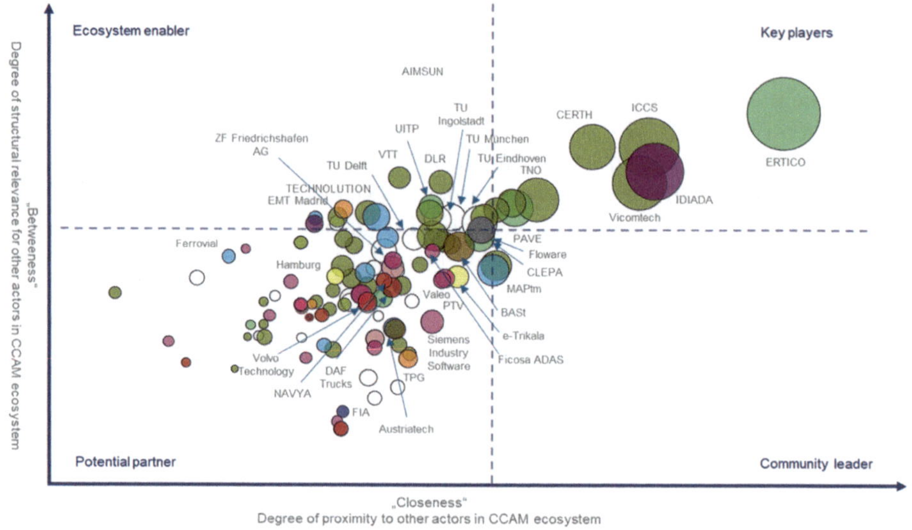

Fig. 3. Latest key stakeholders of the European R&I CCAM ecosystem in Horizon Eu-rope (FP9)

Despite this robust core, certain groups still remain underrepresented in HE-FP9 projects (Fig. 4). These include freight and logistics providers, MaaS operators, local governments, civil society actors, and public transport operators. Their limited involvement indicates a gap in real-world deployment perspectives and user engagement. Furthermore, a notable trend is the decline in participation by OEMs (vehicle manufacturers), which share steadily decreases across the three programmes. This suggests a gradual withdrawal or narrowing engagement from integrators traditionally expected to play a leading role in CCAM deployment. In contrast, organisations classified under "Other Services", particularly those involved in testing, evaluation, and certification, show a modest but consistent increase in participation, pointing to a growing interest in validation and safety assurance processes.

Although coordination projects form a small share of the overall network, they play a central structural role. These initiatives connect distant project clusters, promote knowledge exchange, and support ecosystem-wide alignment—acting as a vital backbone for coherence and collaboration.

In conclusion, the current horizon europe portfolio reflects a strong foundation built on demonstration and integration activities. However, research institutions and universities are still the dominating stakeholders, followed by automotive suppliers and

Table 2. Analysis of research level (HE-FP9).

Top 15 degree	Top 15 closeness	Top 15 betweeness
ERTICO	ERTICO	ERTICO
ICCS	IDIADA	CERTH
IDIADA	ICCS	ICCS
Vicomtech	Vicomtech	IDIADA
CERTH	CERTH	VTT
TNO	TNO	DLR
PAVE Europe	VTI	Vicomtech
TU Eindhoven	PAVE Europe	UITP
MAPtm	TECNALIA	VTI
CRF	Vedecom	TNO
Vedecom	MAPtm	TECNALIA
BASt	Rupprecht Consult	EMT Madrid
CLEPA	FLOWARE	PAVE Europe
VTI	TU Eindhoven	Steinbeis Innovation
UITP	CLEPA	VIF

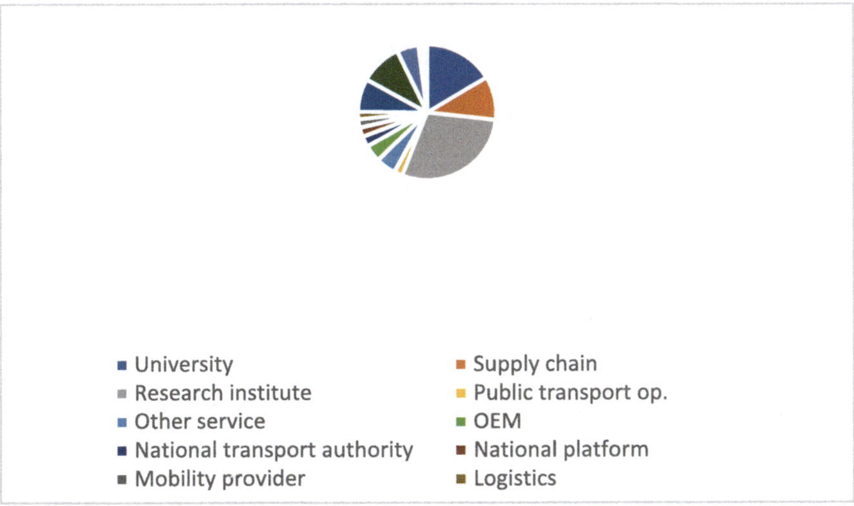

Fig. 4. Distribution of different stakeholder types in the CCAM related R&I projects in Horizon Europe (FP9)

slowly catching up with infrastructure providers and local authorities. To ensure balanced growth and real-world impact, greater emphasis is needed on stakeholder diversity, coordination mechanisms, and underrepresented areas such as societal acceptance, safety validation, and standardisation. Furthermore, the stakeholder ecosystem is built around key players strongly involved in coordination actions and more and more validation activities as well as integrating activities, highlighting a shift from pure technology research towards deployment steps.

Beside domination by few stakeholder categories, we also see a geographical dominance. Countries with a strong automotive and research foundation—particularly Germany, France, and Spain—have consistently led involvement. Greece and Italy has also made notable progress under Horizon Europe, indicating successful national mobilisation around CCAM-related innovation.

However, participation from Eastern European countries remains limited. While there are signs of improvement in the Czech Republic and Poland, engagement from Bulgaria, Romania, the Baltic states, and the Western Balkans continues to be low. This ongoing East–West divide risks excluding important regional perspectives and needs from the development of CCAM solutions.

To further support stakeholder diversification initiatives such as the CCAM SRG and national bodies like the Czech Ministry—are actively working to amplify Eastern European voices and promote broader stakeholder inclusion. Achieving safe, clean, and inclusive mobility across Europe requires a geographically balanced research ecosystem. Expanding participation across all EU member states will not only enhance regional relevance but also strengthen the overall coherence and impact of the CCAM community.

5 National CCAM Ecosystem Analysis with the Example of Germany

German companies are strongly represented in the European R&I funding landscape. Examining national R&I activities provides valuable insight into similarities and differences with the European ecosystem, while also highlighting mismatches and opportunities for better alignment.

An analysis of cooperation patterns in national research projects—focusing on the number of potential networking opportunities—identified the top 10 most networked institutions (Fig. 5). In Germany, research institutes and universities are central to the CCAM ecosystem, closely followed by automotive suppliers. However, OEMs (Original Equipment Manufacturers) are more prominently involved and interconnected than in the broader European network. BASt, as the national transport and road authority, also holds a strong position. In contrast, regional and local authorities, as well as digital and physical infrastructure providers and mobility providers (including public transport operators), are far less integrated.

Overall, research institutes and OEMs form particularly dense networks. There is some collaboration with stakeholders from the infrastructure, insurance, and maintenance clusters, but at a much lower level. Networking with regional authorities is minimal. The City of Frankfurt am Main stands out as a pioneer in this field, participating in numerous projects, followed by Berlin. The first state enterprise to appear in

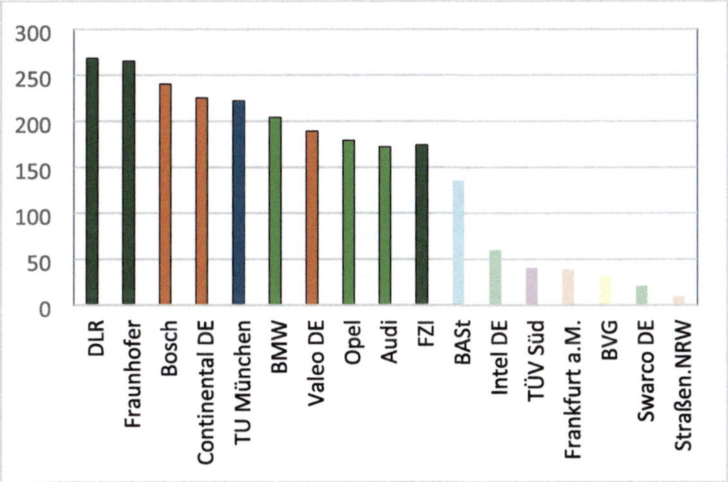

Fig. 5. Key stakeholders of the German CCAM R&I landscape are research institutes & universities, automotive supplier & OEMS, which are complemented in the graph by best connected stakeholder of additional stakeholder categories (e.g. national transport authority, infrastructure, other services, local authority & PTO)

the ranking—Landesbetrieb Straßenbau NRW, responsible for road development and maintenance—has minimal involvement.

The composition of funding recipients reflects extensive cooperation among these institutions. They join forces in various national funding programmes to advance research in technology development, verification and validation, and key enabling technologies such as AI. These consortia often feature repeated participation by OEMs, research institutes, and universities. While this can create redundancies, it also consolidates market power and increases the impact of project outcomes.

From a broader perspective, stakeholder participation in the German CCAM ecosystem mirrors the European picture: research institutes and universities dominate, followed by the automotive supply chain (Fig. 6). Local and regional authorities, as well as infrastructure providers and operators, while not among the top-connected stakeholders, still maintain a notable presence in R&I projects. This suggests that Germany may offer valuable lessons for better integrating these stakeholder types into the European CCAM network.

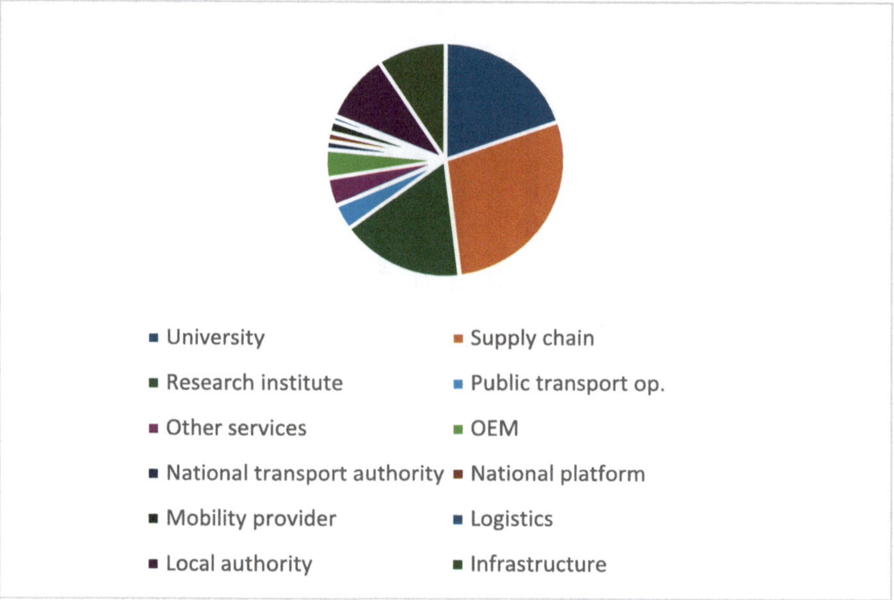

Fig. 6. Distribution of different stakeholder types in the CCAM related R&I projects in Germany

6 Strategic Recommendations, Conclusion and Outlook

To fully unlock the potential of the CCAM ecosystem, Europe must take a set of interconnected actions that strengthen collaboration, inclusivity, and real-world impact.

Strengthening Strategic Brokers and Network Integration
The CCAM landscape has grown into a technologically vibrant network of around 4,000 stakeholders, anchored by highly connected research institutes, cross-sector associations, and major automotive suppliers. These "natural brokers" have the reach to disseminate ideas rapidly and the capacity to connect otherwise fragmented parts of the ecosystem. Empowering them is essential.

Organisations with consistently high closeness and betweenness centrality should be supported to lead cross-cutting initiatives—coordinating standardisation, knowledge dissemination, and systems integration. With adequate mandates and resources, they can accelerate synergies, align standards, and mobilise resources for large-scale demonstrations and deployment.

Activating Associations and City Networks
Large associations and city networks hold untapped potential. Their extensive but often under-connected membership bases could be mobilised to spread best practices, integrate regional clusters into European value chains, and lower entry barriers for smaller or less experienced actors. Achieving this requires moving beyond static membership lists. Structured matchmaking services, shared data spaces, and pre-competitive "safe zones"

for collaboration can enable more dynamic engagement, provided companies are also willing to collaborate outside their traditional silos.

Addressing Underrepresentation and Inclusion Gaps

The analysis reveals persistent underrepresentation of certain stakeholder groups in both research consortia and association cores—particularly public transport operators, freight and logistics companies, MaaS providers, local authorities, civil society actors, and certification bodies. Their limited involvement hinders the translation of technical progress into market-ready, socially accepted solutions. Targeted funding calls, living-lab pilots, and inclusive working groups are needed to better integrate these actors and ensure a more balanced, legitimate innovation process.

Tackling Geographic Disparities and Aligning Levels of Action

Participation from Eastern Europe remains disproportionately low, reflecting a persistent East–West divide. While countries like the Czech Republic and Poland show progress, engagement from the Baltics, Balkans, Bulgaria, and Romania is still limited. Twinning programmes, capacity-building measures, and broader eligibility for demonstration funding could help bridge this gap.

In parallel, better alignment between national and European CCAM initiatives is needed to reduce duplication and strengthen coherence between different funding and policy layers.

Linking Technical Excellence to Societal Readiness

The CCAM community must focus not only on technical breakthroughs but also on deployment-readiness, policy co-design, and public engagement. Aligning R&I efforts with real-world use cases, viable business models, and societal needs will ensure innovations are operationally realistic and socially accepted. This requires embedding regulatory bodies, municipalities, and end users directly into innovation processes.

Positioning within the EU's Industrial Agenda

High-level EU frameworks provide fertile ground for these ecosystem-strengthening actions. The European Commission's Industrial Action Plan for the Automotive Sector—focused on innovation, digitalisation, and autonomous driving—supports large-scale testbed corridors, regulatory sandboxes, and the planned European Connected & Autonomous Vehicle Alliance. This alliance will integrate CCAM with related initiatives such as 2Zero, EUCAR, CLEPA, EFAT, the European Coalition of Local Authorities for Automated Road Transport, and national clusters like SAAM, Austrian SAAM, and Drive Sweden.

Bridge research and deployment through implementation and market readiness

A stronger focus is needed on translating research outputs into operational systems that can be deployed at scale. This requires early integration of implementers such as mobility providers, infrastructure operators, and city administrations into project consortia. Alongside technical development, viable business models must be co-designed to ensure long-term sustainability and commercial uptake. Demonstration projects should include real-world operational scenarios, clear cost–benefit analyses, and pathways for regulatory approval, so that successful pilots can progress seamlessly to permanent services.

Outlook
With a strong knowledge base, maturing structures, and growing cross-sector participation, the European CCAM ecosystem is well placed to become a global leader. By implementing these recommendations—empowering brokers, activating networks, broadening participation, rebalancing priorities, closing geographic gaps, aligning actions, and linking technology to societal needs—Europe can transform its research-rich network into a cohesive, sustainable, and deployment-ready mobility system. A truly inclusive, citizen-centred mobility future is within reach if investment is matched by coordination and commitment across all levels.

Acknowledgements. The authors would also like to express their gratitude to the European Commission for funding this work through the Research and Innovation Action FAME, as well as thank the CCAM Partnership for close collaboration.

Reference

1. FAME Deliverable D3.5: Report on Impact and Outcomes of Stakeholder Engagement

Part III: Users and Human Factors

Being Good (at Driving): Characterizing Behavioral Expectations on Automated and Human Driven Vehicles

Laura Fraade-Blanar[1](✉), Francesca Favaro[1], Johan Engstrom[1], Melissa Cefkin[2], Ryan Best[3], John Lee[4], and Trent Victor[1]

[1] Waymo, LLC, 1600 Amphitheater Parkway, Mountain View, CA, USA
{lblanar,fmfavaro,jengstrom,trentvictor}@waymo.com
[2] Santa Clara University, 500 El Camino Real, Santa Clara, CA, USA
[3] West Virginia University, PO Box 6024, Morgantown, WV, USA
[4] University of Wisconsin-Madison, 500 Lincoln Dr, Madison, WI, USA
jdlee@engr.wisc.edu

Abstract. For over a century, researchers have wrestled with the question of how to define good driving, and the debate has recently surfaced afresh for automated vehicles (AVs). Although numerous principles and models have striven to explore the associated complexities, no framing exists to coordinate and align these concepts into a clear vision. In response, we put forth the concept of Drivership as a framing for the realization of good driving behaviors. Drivership grounds the evaluation of driving behaviors in the alignment between the mutualistic expectations that exist amongst road users. Additionally, we establish a novel vocabulary to more rigorously tackle the conversation on stakeholders' expectations, a key feature of value-sensitive design approaches. We also detail how Drivership comprises safety-centric behaviors, but extends beyond those, to include what we here term socially-aware behaviors (where there are no clear safety stakes).

Keywords: Automated vehicle · driving behavior · ethics · value-based engineering · driving expectations

1 Introduction

On September 13, 1899 in New York City near Central Park, Mr. Henry Hale Bliss alighted from a streetcar and was assisting his companion to exit when he was hit by a car. Because of a truck parked to the right of the lane, the car was driving unusually close to the streetcar. Mr. Bliss died the following day, making him the first U.S. automotive crash fatality. The car's driver was charged with manslaughter and later acquitted as the crash was deemed "unintentional." (Williams 2018; Eschner 2017).

Consider the scene for the 1899 Bliss crash (Fig. 1).[1] The driver's response to a potential conflict (and, possibly, a late reveal) resulted in a negative outcome; yet, depending

[1] Visuals of the crash scenario are not available, so this scene has been reimagined from existing narratives and photographs of period-specific streetcar pick-ups and drop-offs.

G. Meyer and S. Beiker (Eds.): ARTSymposium 2024, LNMOB, pp. 81–105, 2026.
https://doi.org/10.1007/978-3-032-12374-9_6

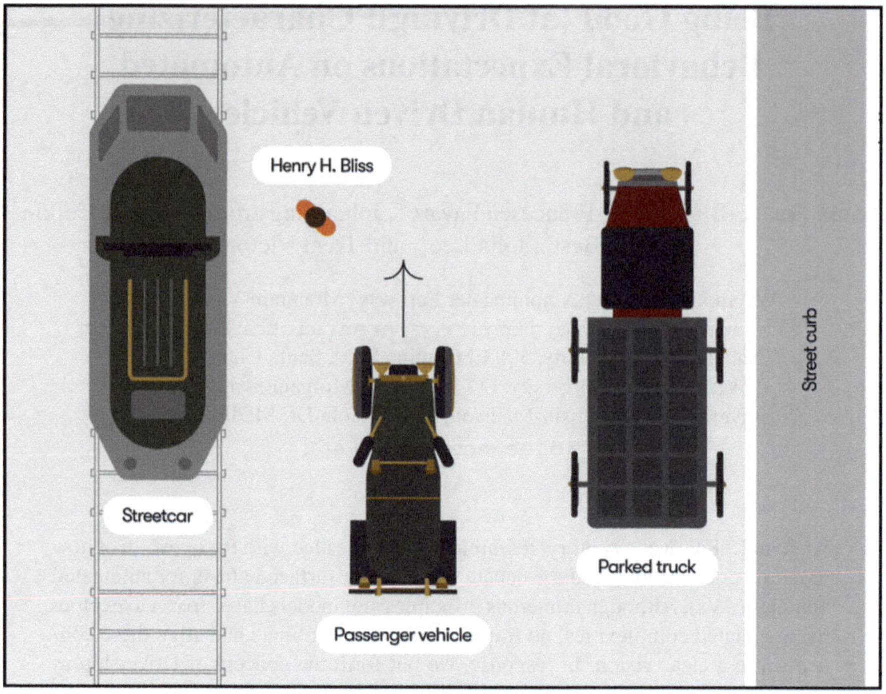

Fig. 1. Hypothesized scenario for the Bliss crash

on the vehicle mechanics, a better response may not have been possible. No traffic laws were broken: New York City did not enact their first traffic code until 1903 (American Safety Council, n.d.). Speeding enforcement was already underway (starting in 1895) with a primary focus on bicycles and a limit of 8 mph, (NYPD, n.d.),[2] but there is no record of the car driving inappropriately fast or recklessly. It could be argued that the driver should have anticipated the risk of passengers stepping out from the streetcar and, hence, should have slowed upon entering the narrow passage in the first place (assuming he did not). The truck's parking choice may have been inconsiderate, given it impacted traffic flow. Mr. Bliss may have unintentionally initiated the conflict by stepping out into oncoming traffic (as a consequence of the streetcar stopping where it did[3]), but he stepped out onto a set streetcar route where road users perhaps should expect exiting pedestrians. Although 125 years have passed since Mr. Bliss's death, this type of scenario and the questions it generates still resonate today. How can the "goodness" of each agent's behavior be defined and evaluated? How can this scenario be understood and

[2] New York City's first speeding law, enacted in 1652, limited horses to a speed less than a gallop (American Safety Council, n.d.).

[3] Although this paper focuses on road user behavior, it is important to note the role that infrastructure plays in influencing behavior. According to contemporary news accounts, "the place where the accident happened is known to the motormen on the streetcar line as 'dangerous stretch,' on account of the many accidents which have occurred during the past summer." ("Fatally Hurt by Automobile," 1899).

used to learn and improve behavior within the transportation ecosystem? And why is this important?

These open questions illustrate the challenge of understanding and assessing good driving behavior of autonomous/automated driving systems and human drivers.[4] To find answers, this paper starts from existing definitions of driving behavior, reviewing the literature to identify different qualifications of goodness and exploring the complexities associated with this determination. In response, we put forth the concept of Drivership,[5] the realization of good driving behavior, predicated on the alignment of driving behavior and societal expectations on driving. Although societal expectations may differ in application between human drivers and automated driving systems, Drivership is driver-agnostic (i.e., applying to both human and non-human drivers). We argue that, within Drivership, identifying and meeting (or exceeding) these expectations entails an analysis that includes safe behavior but also behavior where there are no safety stakes. Furthering this, we disaggregate societal expectations into subtypes through a taxonomic approach, to support further discussion. Finally, we conclude by diving into the question of how Drivership relates to a safety case and a top goal of absence of unreasonable risk.

2 What is Good Driving Behavior?

Considering the purpose of good driving behavior necessitates an ethics-based exploration (Gerdes 2020). Amidst the ethical landscape, a resonating perspective[6] positions good driving behavior as the realization of balancing a communal expectation for road-users to respect each other and be similarly respected when sharing a common road, given consensus on the utility and role of transportation in current society. For example, what responsibility does an ADS developer or fleet operator have to their customers, from one-time users to those who depend on ADSs for their independence and spontaneity, and what is owed to others with whom they share the roadway? What do other road users owe the ADS, given that driving involves public actions occurring in shared spaces? How do disparate goals align to give rise to moral principles on driving behavior which "no one (could) reasonably reject?" (Scanlon 2010). What "mutual recognition" (Ashford and Mulgan, 2007) and mutual constraints should be encouraged or required between road users to provide for a positive shared road experience?

Good driving behavior contributes to roadway citizenship. Roads and adjacent spaces are often public spaces, governed by a specific set of rules and broader societal expectations[7] which, by entering into said space, users agree to abide by. A mutual vulnerability

---

[4] Of note, this paper focuses on driving behavior rather than driving outcomes; leading rather than lagging (Fraade-Blanar et al. 2018). Crashes and deaths are the expression of road users' behavior in conjunction with a range of other factors, such as infrastructure, geography, emergency medical service proximity, etc.

[5] Inspired by the concept of Roadmanship (Fraade-Blanar et al. 2018), Drivership was first presented in 2022 at the Automated Road Transportation Symposium and at Lifesavers, with additional social media (Waymo 2022).

[6] This refers to a perspective which may meet with common agreement.

[7] Societal expectations are discussed in detail in Sect. 3.1.

to each other's behavior on the roadway necessitates this mutual adherence and respect, beyond the general understanding of the applicable rules of the road.

Philosophically-inclined readers will recognize, here and throughout this text, tenets of T.M. Scanlon's philosophy of Contractualism (Scanlon 1998). There is a long history of basing principles of good driving on philosophical tenets: e.g., d'Amato et al. (2022) refer to a Locke-ian (Uzgalis 2001) social contract between road users and with road rules, and it could be argued that Responsibility-Sensitive Safety (RSS) (Shalev-Shwartz et al. 2018) is an example of deontological ethics (Geisslinger et al. 2021; Gerdes 2020). Arguments for a positive risk balance are generally understood to be referencing utilitarianism (Geisslinger et al. 2021). More broadly, it has been argued that general ADS efforts to improve driving behavior are a form of Aristotelian virtue ethics (Geisslinger et al. 2021; Gerdes 2020).

2.1 How Has "Good Driving Behavior" Been Defined, to Date?

Setting aside for the moment the qualifier "good," let's begin with "driving behavior." The technical literature on the sciences of driving and traffic performance[8] is vast. SAE J3164[9] defines driving behaviors as "goal-oriented activities enacted by a vehicle to implement a human driver's or an engaged ADS's (automated driving system) performance of the Dynamic Driving Task (DDT) or DDT fallback." Within current SAE documentation J3016 and J3164 (SAE 2021; SAE 2023), behavior itself has been leveled into (Michon 1985):

- *Operational*: pertaining to the moment-to-moment vehicle control associated with maintaining speed, lateral lane position, and separation between vehicles,
- *Tactical:* pertaining t*o* the ability to make decisions about if and when one should optimally overtake another vehicle, change lanes, select a target speed, etc.,
- *Strategic:* pertaining to broader considerations such as the choice of route and the decision to make a trip.

Several taxonomies have striven to dissect driving behaviors into specific actions or tasks, at times also encompassing safety relevancy and ranking across such tasks. McKnight et al. (1972) identified and then ranked driving actions by criticality, noting that some driver actions are more safety critical than others. The goal was to improve education and generate safer drivers, an effort repeated over the years to similar results (e.g., Bouhsissin et al. 2023).

Let us now turn our focus to what makes driving behaviors "good." The recognition that certain elements of driving behavior are essential for safety represents one of the key ideas behind the concept of *Roadmanship*, introduced in 2018. Defined as a leading measure of safety, it describes "the ability to drive on the road safely without creating hazards and responding well… to the hazards created by others. The concept centers on whether the vehicle 'plays well with others,' even if others are not around," (Fraade-Blanar et al. 2018). The original text included a range of safety-specific metrics such

[8] This report does not aim to be comprehensive of all related literature, but rather notes key texts.

[9] This SAE Information Report is titled J3164 Ontology and Lexicon for Automated Driving System [ADS]-Operated Vehicle Behaviors and Maneuvers in Routine/Normal Operating Scenarios (SAE 2023).

as time-to-collision, maintaining a safety envelope as used in IEEE 2846 (IEEE 2022) and defined in AVSC's Best Practice for Metrics and Methods for Assessing Safety Performance of Automated Driving Systems (ADS) (AVSC 2021), and hard braking, each metric with their own strengths and weaknesses and each drawing from decades of safety research (Fraade-Blanar et al. 2018). Roadmanship also highlighted the importance of conflict avoidance (defensive driving) and secondary crashes[10] (Fraade-Blanar et al. 2018).

The work on Roadmanship refocused a general interest in behavioral safety for the ADS community and ignited new research on the operationalization of technical definitions for the behaviors and maneuvers[11] which directly reflect "good" driving. At the same time, it made clear that good driving behavior included more than just safe driving behavior. To date, much work connecting "good" to specific behaviors or considering behavioral goals focuses on safety or legal compliance with rules of the road. The latter refers to the local and state laws and regulations that determine how drivers and other road users should behave on the road – *what* road users should do, *how* they should do it, and *where* they are when they do it.

There is considerable work on road rules from industry and academia. D'Amato et al. (2022) focused on the aforementioned "social contract for driving," supported by traffic code, driving and case law, etc." Similarly, Xiao et al. (2021) focused on identifying desirable driving from a legal viewpoint. Together these and other works suggest how the traffic code both influences and is influenced by social and ethical expectations. Specifically: "to consider the road rules of the future we start with the rules of the present. Behavior on the road can be seen as governed by rules in four layers: (1) the laws of physics can be seen as constraints on what you can and cannot do; (2) legal rules define what you must and must not do; (3) advisory guidance shapes what you should and should not do; and (4) then there is a layer of social norms and behaviors: what we tend to do and not do. These layers overlap. Many of our social norms align with the rules of the other layers." (Tennant et al., 2021).

Rooted in the second layer, the California Department of Motor Vehicle's regulations for the Deployment of Autonomous Vehicles stipulate the following:

> *[9] The manufacturer shall certify that the autonomous technology is designed to detect and respond to roadway situations in compliance with all provisions of the California Vehicle Code and local regulation applicable to the performance of the dynamic driving task in the vehicle's operational design domain, except when necessary to enhance the safety of the vehicle's occupants and/or other road users.* (CA DMV 2022)

Importantly, this type of regulatory provision recognizes the importance of automated vehicles satisfying the expectations of road rule compliance, while also prioritizing actions for enhancing safety if conflicts between competing road rules (Xiao et al. 2021), or road rules and safety, emerge for that operational design domain.

[10] A secondary crash is a crash that occurs as a result of an original crash either within the crash scene or within the queue or backup in either direction." (Pecheux et al. 2023).

[11] Per SAE J3164, a vehicle maneuver is a "goal-oriented vehicle movement directed by an ADS or human driver in order to achieve a specific result/outcome." (SAE 2023).

Rooted in the third and fourth layer, the approach in ISO 39003 Ethical Considerations Relating to Safety for Autonomous Vehicles is based not on concepts (e.g., safety or compliance with rules of the road) but rather on specific principles:[12] some situational (e.g., Driving Rule DR6: *Make lane changes only if there is an empty 'slot' and there are no dangerous ... consequences,* and DR3: *Even if there are no other vehicles in the vicinity, the vehicle should always be prepared for the sudden appearance of other vehicles with respect to its actions.*), some value-based (e.g., Driving Rule DR4: *During every ride or sortie, show respect and kindness to other road users*, and DR8-a: *Act in a way that acknowledges, communicates with, and respects every entity in the road space.*), etc. (ISO 2023). Unfortunately, taken together, these principles, while laudable, provide little guidance for practical application.

Rooted in the second, third, and fourth layers, the academic literature contains multiple qualifications for good driving behavior, some of which conflict, many of which align, and all of which incentivize slightly-different to very-different behaviors. Among the clearest is the SPRUCE (safe, predictable, reasonable, uniform, comfortable, and explainable) model (De Freitas et al. 2021), and Vinkhuyzen and Cefkin's (2016) social acceptability-focused criteria, describing driving that would "smoothly integrate into the flow of traffic and handle roadway interactions without disrupting other road users." IEEE 7000,[13] in annex G, provides a wide range of ethical values such as care, fairness, politeness, respectfulness, etc. (IEEE 2021) on which one could draw, although not all apply to driving behavior. Other studies used interviews with drivers to conclude that a good driver was characterized by "self-confidence, lack of stress, and not being aggressive," (Doubek et al. 2021), by comfort and naturalness (Peng et al. 2024), or by the interviewee's own personal driving behavior (Roy and Liersch 2013). Reddy et. al. (2024) described a careful and competent driver as one who embodied "awareness, discernment, adaptability, patience, anticipation, caution, roadmanship, responsiveness, communication, and coordination." Still other proposed qualifications for goodness include attributes such as: empathetic (Nordfjærn and Şimşekoğlu 2014), courteous and self-aware (Tillmann and Hobbs 1949), just courteous (Wang et al. 2019), assertive (Oliveira, 2019), trustworthy (Lee and See 2004), compliant (Bin-Nun et al. 2022), fuel-efficient (Wijayasekara et al. 2014), and adaptable to other road user behavior (Schwarting et al. 2019).

There have been some limited attempts in the literature to operationalize individual driving qualities to assess goodness.[14] For example, prioritizing smoothness of movement (Wang et al. 2018), indicating competence and confidence (Alsaid et al. 2020), can be assessed through measuring the minimization of jerk (Itkonen et al. 2017). Or, to prioritize efficiency (Lee and Son 2011), one could assess progress along the route

[12] One of the sources leveraged to build consensus in the ISO technical committee on road safety management [which produced ISO 39003] was the German Ethics Commission on Automated and Connected Driving's report (Fabio et al. 2017), which focuses on good behavior, noting the importance of preventing collisions and, as such, a prioritized approach to safety concerns.

[13] This is the IEEE Standard Model Process for Addressing Ethical Concerns during System Design, a standard for organizations generally [and is not specific to ADS behavior or to ADS developers] (IEEE 2021).

[14] Guidance exists to support the process of operationalizing societal principles and necessary exceptions to specific, value-based actions (e.g., Townsend et al. 2022).

compared to a daily average, reflected in speed choice and how quickly the vehicle negotiates an intersection. Or, a benchmark may focus on defensiveness (Peng et al. 2022), assessed (partially) by the rate of rapid acceleration or deceleration. Comfort could be measured by assessment of vehicle handling and specifically by roll angle (Uys et al. 2006), although evaluating vehicle handling remains a developing field (e.g., (Oh et al. 2021; Hang and Chen 2021)). In the face of this general fragmentation, no clear path arises. Specifically, pegging goodness to driving characteristics is problematic because of the lack of consensus on which characteristics are optimal, and because what is optimal may change based on circumstances and is dependent on context.

A different approach from the enumeration of contributory attributes is that of defining a formal reference for behaviors, known as a behavior reference model. In this approach, the behavior of the driver under evaluation (e.g., human driver or an ADS) is compared against a behavioral model representing some driving performance benchmark.[15,16] Efforts include, but are not limited to:

- **Safety envelope models** establish a formal behavioral model which, given certain assumptions, creates a theoretical guarantee (based on mathematical arguments) that no crashes would happen if the model is universally adopted by all drivers. Examples include Responsibility-Sensitive Safety (RSS) (Shalev-Shwartz et al., 2018), the Safety Force Field (Nister et al. 2019), models based on reachable sets (Althoff & Dolan, 2014; Pek et al. 2020) and/or other spatio-temporal constraints (Favaro et al. 2018);
- **Estimating the momentary risk of collision based on kinematic measures**, for example the Instantaneous Safety Metric (ISM) (Weng et al. 2020);
- **Operationalizing Roadmanship** around specific behaviors such as gap acceptance in roundabouts using naturalistic driving data as a benchmark (Peng et al. 2022);
- **Defining a human driver model** that represents a proper response to an impending hazard. Examples include the Careful and Competent Driver Model (CCDM) (Japan Automobile Manufacturers Association, Inc., 2022) which aims to represent the collision avoidance performance of a "good" (careful and competent) driver in select conflict scenarios, the Fuzzy Safety Model (FSM) (Mattas et al. 2020; 2022) which is intended to represent a driver that "can anticipate the risk of a collision and apply proportionate braking" (ECE 2023), and the NIEON model (Engstrom et al. 2024) intended to represent the collision avoidance response performance of a human driver that is Non-Impaired and have their Eyes ON the conflict. The CCDM and FSM models have been included in UNECE Regulation 157 [R157] on Automated Lane Keeping Systems (ECE 2023).

These efforts vary in usability. Existing implementations tend to be narrowly scoped (e.g., limited generalizability to diverse scenarios), with little to no overarching relationship to other reference models. Moreover, models vary in the quality of signal that they

[15] A potential operationalization that does not use behavior reference models can be found in the digital commentary driving technique, wherein the ADS provides data continuously collected during a trip to create a driving narrative (British Standards Institution 2021).

[16] These concepts are further explored in the in-progress SAE J3330 recommended practice on behavior reference models for evaluating automated driving systems in traffic conflict scenarios (SAE 2024).

provide: the CCDM and FSM were evaluated by Olleja, Markkula and Bärgman (2024) with respect to their ability to predict human collision avoidance performance in the SHRP2 naturalistic driving dataset, testing the assumption that they properly represent a "careful and competent driver." For both models, it was found that the collision behavior they exhibited deviated significantly from that of human drivers, where CCDM generally responded to the hazard later than human drivers while the FSM responded earlier.

While a relatively large number of usable behavior reference models and other operationalizations of good driving behavior exists, an overarching framework is needed to develop consensus to guide operationalization and facilitate alignment on behavioral benchmarks (IEEE 2024). Although the literature shows many ways to describe driving behavior, to describe goodness, and to measure driving behavior and compare it to a benchmark representing goodness, no framing exists to coordinate and align these concepts into a clear, cohesive vision (Reddy et al. 2024), in order to provide precise guidance on operationalization.

We propose a novel framing, grounding good driving behavior in realizing societal expectations. This framing is called Drivership. In this context, Drivership[17] becomes an expression of how mutualistic expectations between all road users are being met by drivers on the road.

3 What is Drivership?

The concept of Drivership[18,19] has been developed as a framing of the realization of good driving behavior. For ADSs, company-internal development and evaluation, shaped by an array of endogenous and exogenous factors, beget the behavior the ADS will exhibit on the road. Drivership research advances the ongoing process of aligning driving behavior and societal expectations,[20] engendering public trust and societal acceptance (Fig. 2). Particular attention to alignment may be necessary as societal expectations and behavior change, e.g., in reaction to new technology.

Safety lives at the core of Drivership, building on the aforementioned concept of Roadmanship (Fraade-Blanar et al. 2018). Roadmanship, which focuses only on safe driving behavior, and numerous other sources make clear that driving safely is a societal

---

[17] In this document, Drivership is primarily discussed from a passenger vehicle driver viewpoint, but that is not to say this is the only entity on which expectations are placed. There are societal expectations on the behaviors of pedestrians, bicyclists, truck drivers, etc., some of which are enacted into law.

[18] Drivership operates across all levels of driving behavior (Michon 1985). Notably, at the tactical and operational level, driving behavior is constrained by the laws of physics; what is physically and technologically feasible.

[19] Drivership spans the pre-crash driving behaviors, and crash and post-crash behaviors [should a crash occur], drawing from the Haddon Matrix (US DOT 2011). For example, maintaining proper motion control is essential to prevent a crash, and essential during and after a crash has occurred to minimize further harm.

[20] An important role of assessing Drivership is to surface conflicts between different social expectations and stakeholder expectations, and to guide resolution. Formalizing such tradeoffs could identify pathways to mitigation (2012).

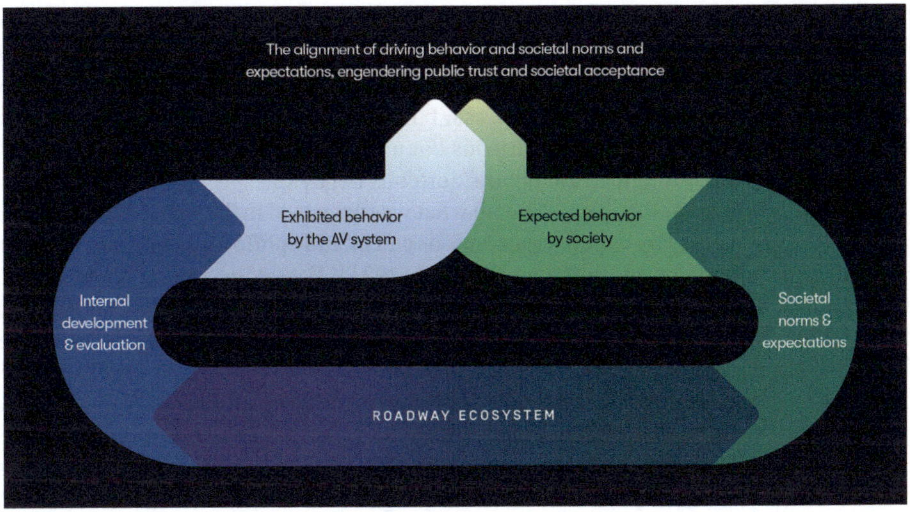

Fig. 2. The alignment of driving behavior and societal norms and expectations for ADSs[21]

expectation for road users. Focusing on expectations of behavior that clearly and directly relate to safety is, however, *necessary but not sufficient* for the full realization of *"good"* driving behavior. Standards such as IEEE 7000 (2021) and publications described in Sect. 2.1 above, make clear that expectations for the realization of good driving behavior extend beyond safety.

Consequently, we position Drivership as ranging from *safety-centric* to *socially-aware* behavioral expectations.[22] Safety-centric refers to expectations on behaviors which directly influence the probability of an injury. In a causal chain (as described by Favaro et al. (2018)), these are hazardous behaviors immediately proximal to harm. Conversely, socially-aware behaviors are more distal to harm manifestation, and without a clear or proximal injury risk. These behaviors may be more related to expectations around the social customs of driving. For example, a trajectory or proximity conflict

[21] Social expectations may be deeply or lightly rooted in a given locality. The variation between localities and the likelihood of an expectation changing is an important topic for future discussion.

[22] Drivership draws from parallel concepts in other transportation modes, such as seamanship and airmanship. These contexts have a more lengthy and rich history of skilled vehicle operation and strictly apply to trained professionals. Additionally, the concepts of seamanship and airmanship are still somewhat nebulous and lack an agreed upon definition [e.g., (Kongsvik et al. 2020; Nergard 2014). Though resistant to a precise definition, consistent themes of skilled operation in the context of the air and sea apply directly to the concept of Drivership. The base definition of both builds upon the precept that a thorough understanding of the rules of operation are a necessary but insufficient factor in establishing expertise. Airmanship and seamanship are communicated and demonstrated through action. Safe, reliable, and successful actions in routine trips serve to demonstrate understanding of the rules of operation, but similar actions in atypical scenarios communicate an important aspect of skilled operation – the ability to act in ways incongruent with rules, if and when necessary, to avoid negative outcomes.

(e.g., the subjective perception of cutting people off, or a pedestrian passing too closely) may be interpreted as uncourteous or normal depending on local social customs.[23] In describing driving behaviors, note that behaviors aligned with Drivership can have safety considerations but no social considerations, but not vice versa. In other words, Drivership cannot be realized through a courteous but unsafe behavior, but it remains possible to attain Drivership through a safe but uncourteous behavior.

The notion of Drivership is in line with what was proposed in the IEEE 7000 standard, which details processes by which organizations can include philosophical considerations in product development and refinement (IEEE 2021). By taking into account societal expectations from trip start through trip end, there is opportunity to consciously and deliberately align with value-sensitive system design methodologies[24] (Friedman et al. 2015). This standard, which is not specific to the transportation or technology industries, does not prescribe specific ethical tenets but rather focuses on an Ethical Values Elicitation and Prioritization process involving stakeholder groups. The standard notes within the definition of a stakeholder that "some stakeholders can have interests that oppose each other or oppose the system." Additionally, "groups of stakeholders: internal, users, opponents, and external authorities, are treated differently when risks, ethical values, and impacts are evaluated" (IEEE 2021).

3.1 Exploring Societal Expectations Around Good Driving Behavior

Societal expectations can be explored through an established (and much debated) thought experiment. Philosopher John Rawls asked what rules a group of individuals would make if they did not know each of their own "place in society, class position, wealth, or social status, nor does anyone know their race, gender, fortune or misfortune in the distribution of natural assets and abilities, level of intelligence, strength, education, and the like." (Freeman 2008). He argues that in such a situation, the principles that all individuals rationally agree to (or not reject) are fair, and that fairness extends to any institutions, laws, and conventions required by these principles. Consequently "the role of the social contract is to represent this idea, that the basic principles of social cooperation are justifiable hence acceptable (or not rejectable) to all reasonable and rational members of society, and that they are principles which all can commit themselves to support and comply with." (Freeman 2008).

There is an interesting and real parallel in the transportation ecosystem, wherein we accept risks as a society but experience them as individuals (Tingvall and Lie 2021). The average road user regularly plays multiple roles, moving from being a driver or a passenger in a car or bus, to a pedestrian (even if just walking from the transit stop, car park, or bicycle rack to their destination), to a cyclist, etc. often in the same week.

[23] What is considered acceptable and expected will differ depending on local driving culture. For example, what is considered an unexpected or aggressive "cut-off" during a lane change in one region, may be expected and necessary in another region.

[24] Value sensitive design is an approach to engineering design calling for the inclusion of broad stakeholders' perspectives within the conceptualization of a system design, beyond those involved in the traditional development value-chain, to include those who will be impacted by the existence of a product [Himma & Tavani, 2008; Friedman et al. 1996].

Norms and expectations are generated by a collective, consensus understanding of many road users, each of whom uses the road in different ways at different times and who may shortly be occupying a different role.[25] Meeting said expectations becomes the realization of the social contract between road users. Consequently, societal norms and expectations, by their very nature, include the mutual agreement on acceptable driving behavior.

Recent work in social philosophy distinguishes two types of expectations[26,27]:

- Empirical expectations: "beliefs that a certain behavior will be followed." (Bicchieri et al. 2023)
- Normative expectations: "beliefs that a certain behavior ought[28] to be followed." (Bicchieri et al. 2023)[29]

The source of these beliefs is not in scope for the present paper, but generally it is attributed that past experiences shape Empirical Expectations,[30] and agreed-upon principles shape Normative Expectations.[31,32] Additionally, Empirical Expectations are grounded in descriptive norms, while Normative Expectations are grounded in social norms (Bicchieri 2006). As the reader may have gleaned from the definitions, these two sets are not disjointed. In fact, as presented below, Empirical Expectations can, over time, influence Normative ones. Additionally, Normative and Empirical Expectations can result in the same behavior (i.e., when what one has previously done is what one should do).

---

[25] For example, as we get out of a car, and walk a block to our destination, we transition from passenger to pedestrian.

[26] This is not to judge one type of expectation as being superior or preferable to another, but rather to draw a taxonomic set of distinctions between them, to aid discussion.

[27] Expectations are rooted in context; how we as a society form and seek the fulfillment of expectations around driving differs compared to flying or taking a train (Tingvall and Lie 2021).

[28] "Ought" is used by Bicchieri to focus on expectations drawing on mutual respect; Bicchieri asks us to "consider… a rule of equal division. In this case we may believe that others ought to 'divide the cake in equal parts' because this is the fair thing to do. We think they have an obligation to follow the rule, a duty to be fair." (Bicchieri 2006). Please note, a legal definition of duty is not being used by the author being quoted here.

[29] Normative Expectations have clear connections to the concepts set forth in IEEE 7000. For example, the standard uses the term personal maxims to describe a "personal principle of what one wishes for, acts upon, and thinks that it should be applicable to everyone." (IEEE 2021).

[30] These expectations draw on past experiences, such that "one expects people to follow R in situations of type S because one has observed them to do just that over a long period of time. If the present situation is of type S, one can reasonably infer that, ceteris paribus, people will conform to R as they always did in the past." (Bicchieri 2006).

[31] Normative expectations can be elicited from a wide range of sources, including the technical, peer-reviewed work and grey literature, rules of the road, regulations, training materials for new drivers, but also through structured interviews, casual conversations, conference presentations, etc. There is no one source for Normative Expectations.

[32] Empirical and Normative Expectations [confined by the laws of physics and technological capabilities] may vary based on stakeholder role and personal viewpoint [Pyrialakou et al. 2020; Othman 2021; Yuen et al. 2020; Nair and Bhat 2021].

Importantly, "the Normative expectations condition says that expectations are believed to be reciprocal. That is, not only do I expect others to conform, but I also believe they expect me to conform… In large, anonymous groups (as occurs on the roadway), if we do not want to act contrary to others' normative expectations it must be because we find such expectations reasonable" (Bicchieri 2006). How the range of Normative Expectations on ADSs may differ from expectations for human drivers and human drivers supported by ADAS is an ongoing, industry-wide discussion (e.g., (Bonnefon et al. 2024; Hulse et al. 2018; De Freitas et al. 2021)). Additionally, Normative and Empirical Expectations on behaviors change, stemming from shifts both in the transportation ecosystem (e.g., the introduction of shared e-scooters) and beyond (e.g., changes in distracted driving rates due to cell phone usage).

Examples of discrepancies between these Expectations are rife in traffic safety. For example, it is an Empirical Expectation that some drivers are drunk (according to IIHS, 30% of all fatally injured drivers had a BAC > 0.08% in 2022 (IIHS, 2024a)), whereas it is a Normative Expectation that road users ought not to be drunk. Similarly, it is an Empirical Expectation that drivers speed (according to IIHS, there were 12,000 deaths in speed-related crashes in 2022 (IIHS, 2024b)); it is a Normative expectation to drive at a safe speed, anchored by the posted limit. Car seat usage for infants is an example of a behavior where the discrepancy between Empirical and Normative Expectation is now relatively small; according to a 2021 study, 98.8% of children under the age of 1 were in a car seat (Boyle 2023). Minimizing discrepancies such that Empirical Expectations align with Normative Expectations supports improvements on the roadway. Furthermore, calibrating ADS behavior to support this minimization is the work of internal development and evaluation (a process outlined in Fig. 2).

Empirical and Normative Expectations do not directly parallel the distinction between unsafe and safe behaviors. For example, well-intentioned stakeholders might have Normative Expectations, which if followed, could have unintended side-effects with negative safety repercussions (Fraade-Blanar et al. 2018). Consider, for instance, a situation in which hard brake rates dictate insurance premiums, inadvertently setting what could be perceived as a Normative Expectation that one should not hard brake. One could thus feel incentivized to limit hard brakes. However, research shows that hard braking is a common mechanism to avoid collisions; disincentivizing it may have unintended negative consequences.[33] Nuance, in general and specifically around possible unintended consequences, is required in considering the relationship between Normative Expectations and realized driving behavior. Additionally, it is possible for there to be

[33] As noted in Fraade-Blanar et al., (2018) recording of these events "would unfairly penalize the driver taking evasive maneuvers to avoid a crash" and could, at an extreme, disincentivize said avoidance. Guidance on how to proceed in such circumstances is limited; for example, the aforementioned ISO 39003 DR8-b notes the importance of acting in a manner that is 'optically' correct but, despite noting checks for internal consistency, fails to suggest how to proceed when this conflicts with what is safest (ISO 2023).

multiple beliefs about a given behavior within a broad society, when there is no settled, dominant Normative Expectation (Bicchieri 2006).[34,35]

Beyond expectations of what will happen (based on past experience) and ought to happen (based on societal obligation), there is a third type of expectation:

4 **Furtherance Expectations:** beliefs on which behavior(s) could be exhibited to enable continuous improvement of the transportation ecosystem, motivated by the intention to improve the status quo.

This type of Expectation stems from Drivership's contribution to roadway citizenship. It refers to expectations as opportunities to improve the road traffic system in terms of safety, efficiency, comfort, ease, etc. which could eventually be established as Normative Expectations in a society.[36]

All three types of expectations are constrained by what is possible and technologically/logistically/physically/etc. viable. It may be that for a given topic, Empirical, Normative, and Furtherance Expectations align or diverge (Fig. 3). For example, if a low-severity collision occurs on a freeway, it may be an Empirical Expectation that the collision counterpart drives off (i.e. a hit-and-run) and the other party remains in place where the collision occurred. It may be a Normative Expectation that both parties pull off to the shoulder. However, researchers and law enforcement (e.g., Peiris and Harik 2018; Texas Transportation Code 2005; AZ DPS 2022; ABC 30 Action News 2021)) may consider that that it would cause less congestion and provide a safer option if both parties exited the freeway, if possible, and then pull over to reduce the chances secondary crash (either involving the original vehicle and another vehicle that comes upon the scene, or other vehicles who are maneuvering to avoid the scene). This behavior lives in the overlap between Furtherance and Normative Expectations; choosing to enact this last behavior represents the ongoing process of modifying expectations towards what is established as preferable based on mutual responsibility to all road users on the roadway, bearing in mind what is possible (e.g., the vehicles were drivable and there is an exit nearby).

[34] For example, a prevailing Normative Expectations has not been established around hands-free cell phone use while driving; specifically, if talking on a handsfree device while driving is acceptable (the same survey found 19% believed it was dangerous and 42% supported a law prohibiting the behavior) (NCSL 2024).

[35] Drivership specifically refers to Normative Expectations "that are followed by a majority of the target population, significantly affect behavior (intuitively, 'strong' norms), and do not fade over time without external intervention (intuitively, 'stable' norms)." (Bicchieri and Demo 2024).

[36] What is involved in fulfilling an Expectation varies. This variation does not result in prioritization of expectations or denote quality or realism. Instead it speaks to what it means to achieve or exceed a given expectation and, by extension, an organization's role in said achievement. For example, for Furtherance Expectations like those inspired by Vision Zero, driving behavior is only a portion of the overarching concept which includes roadway design, vehicle design, etc. (Lie and Tingvall 2024; Ecola et al. 2018). Vision Zero cannot be achieved through the efforts of any one company or organization as it entails shared responsibility. But a contribution to Vision Zero may be shown through, for example, a safety record, evidence of considering impact on the roadway as could be assessed by the FIA Foundation Road Safety Index, participating in safety-focused standards and best practices bodies, etc.

Societal movements, regulatory efforts to shape/revise/enforce rules of the road, advocacy, etc. aim to pull Empirical Expectations to overlap with Normative Expectations by normalizing and then entrenching the desired behavior.[37] Careful, deliberate research and cross-functional discussion represent the process of pulling Normative Expectations into closer alignment with Furtherance Expectations. These three Expectations and what pulls them closer to or prevents them from aligning represents important topics for future research.

As society naturally shifts over time, expectations change. This includes changing conceptualizations on what will happen (Empirical), what should happen (Normative), and what could happen to improve the transportation ecosystem (Furtherance). Tensions may exist across the categories of Expectation (intuitively represented in Fig. 3). We argue that Drivership operationalization is grounded in Normative Expectations that are shared at a societal level.[38] These Drivership Expectations have **Feasible Constraints**; regardless of their basis and/or source, these expectations are confined to what is currently technologically and physically (i.e., according to the laws of physics) achievable and practical.[39]

A sound conceptualization of Drivership thus needs to ensure that the attributes through which it is evaluated reflect current, societal, feasible Normative Expectations. Operationalization occurs with an understanding that non-Drivership expectations may exist (i.e., non-Normative Expectations, Normative Expectations that are beyond the Feasibility Constraints, Expectations that do not exist at the societal levels, etc.) and that the ADS is only one of many road users.

3.2 Towards the Operationalization of Drivership

The operationalization of Drivership[40] may require a range of different methodologies, including leveraging expert judgment, in combination, where available, with quantitative

[37] For example, expectations around seat belt use (and subsequently, rates of seat belt use itself) changed as a result of the 70% by '92 program (CDC 1992), the subsequent Click it or Ticket Campaign, etc. (Solomon et al. 2004; Nichols and Solomon 2013).

[38] This situation of Drivership Expectations at the societal level clearly recalls its roots in Contractualism; focusing a "set of principles for the general regulation of behavior that no one could reasonably reject as a basis for informed, unforced, general agreement," (Scanlon 1998) and motivated by a combination of "self-regard and by respect for others." (Ashford and Mulgan 2007).

[39] An excellent illustration of Feasible Constraints can be found in stopping distances. In 2021, according to Consumer Reports, the average stopping distance (from 60 mph) for passenger vehicles was 132 feet (Bartlett 2021). A more recent article noted the Ford Mustang Dark Horse set a new record with a stopping distance (from 60 mph) of 86 feet (which involved an average declaration of 1.4 Gs) (Dushyant 2023). Consequently, an expectation of a distance of 75 feet between an obstacle being introduced and the vehicle coming to a full stop from 60 mph would be outside of the Feasible Constraints, because it is not technologically possible currently. An expectation of 125 feet would be inside the Feasibility Constraint, while a stopping distance of 175 feet would not be Normative.

[40] The evaluation of Drivership is highly context dependent and involves a holistic understanding of the spectrum of behaviors that each traffic participant could have undertaken and the entirety of the scenario timeline. In evaluating, it may prove helpful to speak of Micro Drivership

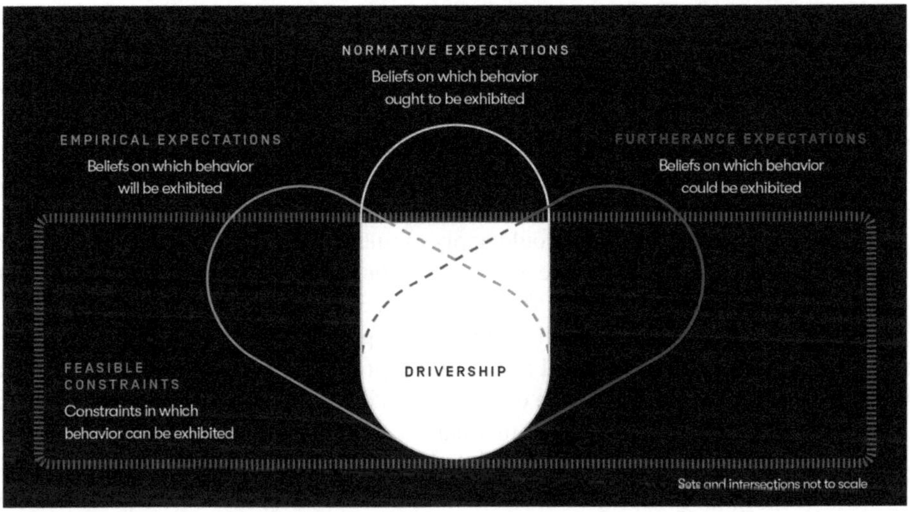

Fig. 3. Drivership (in purple) in relationship to societal Normative, Empirical, and Furtherance Expectations

behavior reference models (like those listed in Sect. 2.1). These models have benchmarks calibrated to or based on driving qualities that meet societal expectations.

One example of such a behavior reference model, applicable to the expectation of responding properly to potential and actual conflicts, is the *NIEON model.* This model is intended to represent the collision avoidance response performance expected from a Non-impaired human driver with their Eyes ON the conflict. Engström et al. (2024) describes how such a quantitative response model can be fitted to human data from naturalistic driving that fulfill the NIEON criterion and Scanlon et al. (2022) demonstrate how it can be used as a benchmark in simulated collision avoidance testing of ADSs. The NIEON model is thus intended to represent Normative Expectations on collision avoidance performance, as a benchmark for ADS performance. This can be contrasted to a collision avoidance driver model designed to represent Empirical Expectations, that is, how drivers in the general population (some of which may be distracted, impaired, etc.) are expected to respond to conflicts. Such a model may, for example, provide a probabilistic representation of the collision avoidance performance of the general driver population (for example, Bärgman et al. 2015; Engström et al. 2023).

Another example of a reference model is Surprise, described in Dinparastdjadid et al. (2023). This is a novel type of model applicable to evaluating the expectation of avoiding initiating conflicts. More specifically, the Surprise model can be used to

(i.e., analysis of a specific road user maneuver) and Macro Drivership (i.e., analysis of the chains of maneuvers by different road users, bearing in mind the broader goals and all the alternatives available to the different road users, long-term implications, counterfactuals, etc.). It may be that behavior within a scenario, when taken more broadly, contributes to a larger goal of safety specifically and goodness broadly, but one specific maneuver appears to be poor (Micro) Drivership. Essentially, "the wrongness of an action depends on the circumstances in which it is performed." (Scanlon 2010).

evaluate the degree of surprise associated with different maneuvers such as braking and lane changing, as experienced by other road users. The model is based on comparing the predicted behavior of the ego vehicle to the actual behavior (i.e., probabilistic mismatch surprise) or, alternatively, how much a current belief about the future behavior of the ego vehicle changes with new observations (i.e., belief mismatch surprise). Since traffic conflicts are typically initiated by behaviors that surprise other road users, avoiding surprising behaviors is key to avoid initiating conflicts.

Other types of benchmarking could apply to rules of the road's compliance expectations. The evaluation of compliance with rules of the road remains among the fundamental tenets of Drivership; within the framing of Fig. 3, one could explore setting internal benchmarks to evaluate the compliance-focused appropriateness of automated vehicle behaviors. On one hand, Empirical Expectations grounded in the observation of pervasive violations on the part of human drivers would likely offer low rigor for establishing an acceptable benchmark.[41] On the other hand, Furtherance Expectations pushing for a perfect record based on hard coding of individual rules may discount emergency situations in which rules are violated to preserve safety or in response to instructions from officers on the road.[42] A composite benchmark could thus be devised to combine the verification of ability and proficiency (for example, through scenario-based testing) with appropriate product governance and management processes based on internal monitoring of on-road operations.[43] These processes could be informed by the setting of feasible (recalling feasibility constraints) and acceptably low rates of occurrence of pre-specified events of interest. Reference models could then support the devising of behavioral criteria for compliance capabilities: for example, evaluating timing decisions to proceed at a yellow light, or stopping location in relation to the geo-specific normative stop line.

Figure 4 visualizes the role that behavior reference models can play within the evaluation of Drivership, and the connection to societal expectations. This figure shows how the socially aware and safety-centric qualities of driving behavior are assessed, by individual or overlapping behavior reference models. On the right is a taxonomic disaggregation of Drivership, the discrete components of which can, in turn, be evaluated through the use of behavior reference models (at the bottom of the visual). On the left is the flow illustrating how Empirical, Normative, and Furtherance Expectations reciprocally relate to this decomposition. Of note, the operationalization of Drivership - as grounded in feasible Normative Expectations - involves the precise definition of behavioral requirements for the ADS derived from such Expectations. Conversely, the specification of behavioral reference models stems from such behavioral requirements.

[41] Of note, in actual practice, there are many reasons human road users may make an intentional road rule violation (which is distinct from an (unintentional) mistake or slip (Parker et al. 2007; Reason 1990)). As an extreme example, a survey of bicyclists in Montreal noted that only 0.6% chose behavior that consistently followed traffic laws, with "it was the safest option," or "it made sure I was visible to cars" as the most common reason for not following traffic laws (Chaloux and Ahmed 2019).

[42] See, for example, the CA DMV provision of Article 3.7 §227.32 [c], stating "except when necessary for the safety of vehicle's occupants and/or other road users." (CA DMV 2022).

[43] This is aligned with the approach Waymo presented in its 2020 paper overviewing the role of various methodologies for safety readiness determination (Webb et al. 2020).

Two-way arrows in the figure denote the mutualistic nature, as previously presented in Fig. 2.

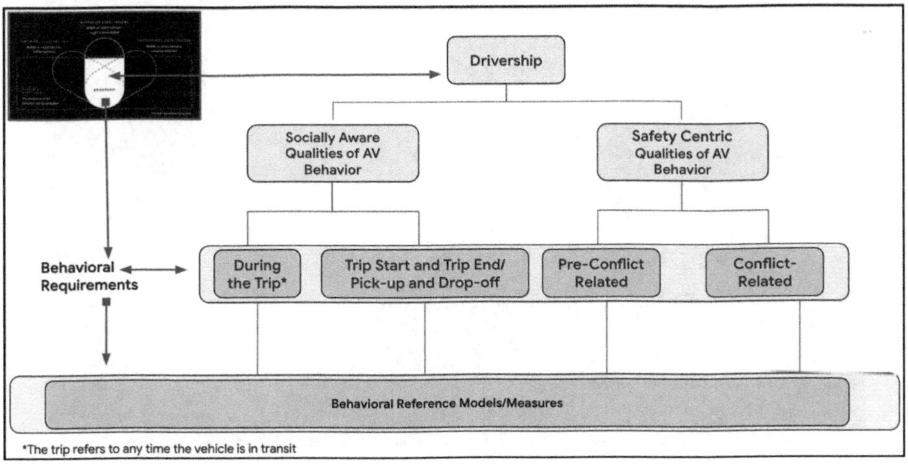

Fig. 4. A broad framework of the operationalization of Drivership, in connection with Expectations

4 What is the Relationship Between Drivership and a Safety Case?

Having established that Drivership embeds both considerations of safety and social awareness, the question then becomes what is the connection between Drivership and a safety case. A safety case is a "structured argument, supported by a body of evidence that provides a compelling, comprehensible, and valid case that a system is safe for a given application in a given environment" (UL 4600 2023). Based on the standardized definition of "safety" (ISO 2018; ISO 2022; Favarò et al. 2023), the determination of Absence of Unreasonable Risk (AUR)[44] is established as the top-level goal. Determining Absence of Unreasonable Risk involves assessing risk from behavioral, operational, and architectural hazards[45] (Favarò et al. 2023). For behavioral hazards specifically, Drivership provides the means to assess reasonableness of behavioral risk in connection to valid societal moral concepts (ISO 2018, 2022), denoted here as feasible Normative Expectations. Consequently, Drivership supports identification of the types of driving events most relevant for the behavioral component of Absence of Unreasonable Risk (AUR). Figure 5 visualizes the intersection between improper Drivership and risk of harmful outcomes in interactions between road users.

44 An unreasonable risk is a "risk judged to be unacceptable in a certain context according to valid societal moral concepts." (ISO 2018).

45 Per ISO 26262, a hazard is a "potential source of harm" (ISO 2018). Within the context of Drivership, we focus on behavioral hazards as those associated with potential sources of harm resulting from the ADS's displayed driving behavior. (Favaro et al. 2018).

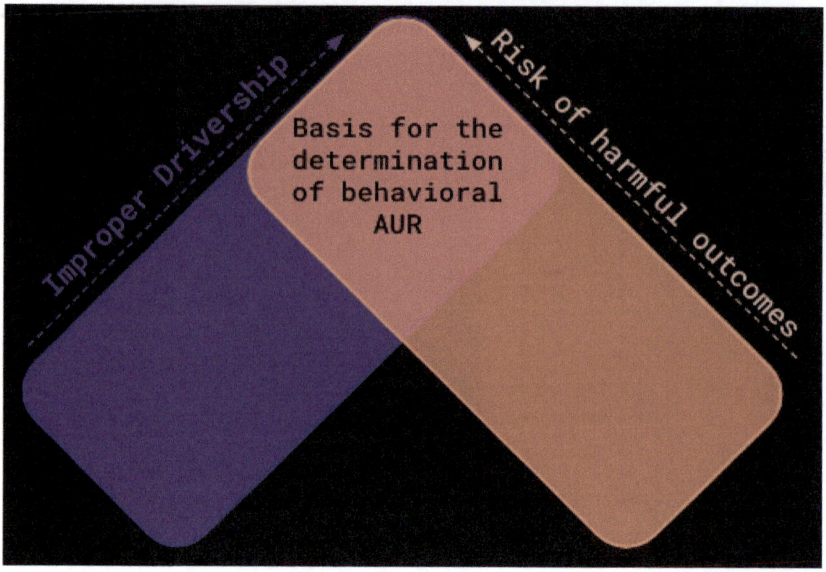

Fig. 5. The intersection of improper Drivership and the risk of harmful outcomes

While behavior involving improper Drivership (represented in lavender in Fig. 5) and behavior where there is a risk of harmful outcomes (represented in grey in Fig. 5) should be continuously analyzed as part of engineering improvement practices, they should be appropriately differentiated. Specific to behavioral AUR, it is important to differentiate interactions where (A) there is improper Drivership but no risk of harmful outcomes, from (B) the behavior does not represent improper Drivership but, as a result of the larger interaction in which it occurs, it carries a safety risk, from © there is improper Drivership involving a risk of harmful outcomes.

Drivership's inclusion of safety-centric and socially-aware behavior can add specificity to safety evaluation practices along two directions:

1. Risk can still be regarded as *reasonable* if the driver exhibits proper (even if non-ideal) Drivership, and even if a high severity collision results. Evaluation provides a needed lens to identify events wherein there is both a risk of harmful outcomes and improper Drivership; this lens allows appropriate prioritization safety-critical events, differentiating, for example, the relevance of conflicts initiated by other road users which still showcase proper Drivership by the responder (e.g., their proper response resulted in the successful mitigation of outcomes to a lower degree of severity (as illustrated by (B) in Fig. 5)).
2. Improper Drivership, which results from a misalignment between driving behavior and Normative Expectations (as illustrated in Figs. 2 and 3), may not, on its own, be a sufficient tell of a safety concern. This is nested in the very definition of socially-aware attributes of Drivership (as described in Sect. 3. *What is Drivership?*), which are more distal to harm manifestation and without a clear or proximal injury risk.

Behavior reference models and methodologies used to evaluate Drivership contribute to the evaluation of Absence of Unreasonable Risk within a safety case, by assessing the relationship between on-road behavioral hazards and societal expectations. Thus, evaluating Drivership is key to assessing the extent to which the risk associated with a given behavioral hazard is unreasonable[46] (Favarò et al. 2023). For Drivership, the evaluation of a specific event would go beyond the contribution to a safety determination to consider appropriateness of and societal expectations around the behavior.

Finally, because Drivership includes but is not limited to safety, it is considered in all stages of the Safety Determination Lifecycle (Favarò et al. 2023; AVSC 2024). Herein, Drivership serves to inform product behavioral development cycles, as well as the appropriate setup of coherent internal processes and methodologies to aid credibility and confidence into the performance evaluation as it may contribute to a safety case.

5 Conclusion: How to Move Forward

A key question motivates Drivership: what is good driving behavior? This question concerns not just ADSs but rather all involved in the field of motorized transportation, which has been enmeshed in society for over a century. It is intuitively clear that there is no objective answer to this question. Our survey of the subjective answers of other researchers and policymakers unearthed a wide variety of topics, values, principles, and models, without either a clear path to operationalization nor an overarching framing. Additionally, these answers frequently lacked a clear justification for the inclusion or exclusion of certain attributes or perspectives.

We thus began by refining the question on good driving behavior to: what do road users owe each other (Scanlon 1998) This iteration, in turn, led us to analyze stakeholders' expectations, where, based upon prior literature, we taxonomically distinguished between: Empirical, Normative, and Furtherance Expectations. We chose to anchor Drivership in Normative Expectations at the societal level, limited by what is feasible; **in the simplest of terms, good driving behavior as the behavior that a driver ought to do and plausibly can do, to be a good citizen of the roadway (feasible Normative Expectations).** This may or may not overlap with the behavior they are anticipated to do (Empirical Expectations), and/or could do in order to improve the transportation ecosystem (Furtherance Expectations).

We offer a path toward operationalization by discussing how, among other tools, Normative Expectations can be used as the basis for the development of benchmarks implemented by behavior reference models. We also sought to distinguish qualities of goodness in driving behavior; safe, of course, but also socially aware. This distinction is important when connecting Drivership to a safety case, which requires in-depth evaluation of those events that combine both improper Drivership and potential for injury risk.

Drivership is a realization of good driving behavior, providing a pathway to achieving this goal. Growing and maintaining a Drivership approach includes:

1. Developing and supporting the architecture and methodologies that contribute to a determination of Drivership;

[46] Please note, a legal definition of reasonableness or unreasonableness is not being used here.

2. Generating and supporting an evolving understanding of dynamic Normative Expectations with the public, policy-makers, and other key stakeholders to support operationalization;
3. Analyzing events to understand alignment between behavior and societal expectations;
4. Planning for continuous Drivership improvement to be crafted, according to realistic expectations on engineering capabilities and technical challenges and constraints;
5. Across stakeholders (and beyond developers), engaging in consensus-creating-and-confirming activities regarding Normative Expectations on ADS behavior.

This represents a cyclic process that supports a value-based system approach. Understanding Empirical, Normative, and Furtherance Expectations represents not a one-time occurrence but an ongoing, progressive process of change toward the world we want to live in; evolving answers to the question of good driving behavior requires regular re-alignment into the future. Further research is needed to develop Drivership into a technology-, and autonomy-neutral approach to understanding good driving behavior. Advancing a mutualistic Drivership framing provides definition and supports evaluating "goodness" of behavior, from the road users involved in Mr. Bliss's death in New York 125 years ago to those seen today, and into tomorrow.

References

ABC 30 Action News: Know the Road with the CHP: What to do when you have car problems or are in a crash on the freeway. ABC 30 Action News, 28 June 2021. https://abc30.com/driving-road-safety-chp-california-highway-patrol/10840461/

Alsaid, A., Lee, J.D., Price, M.: Moving into the loop: an investigation of drivers' steering behavior in highly automated vehicles. Hum. Fact. **62**(4), 671–683 (2020)

Althoff, M., Dola, J.M.: Online verification of automated road vehicles using reachability analysis. IEEE Trans. Robot. **30**(4) (2014)

American Safety Council: The History of Speed Limits in America: A Nation Speeding Up. American Safety Council (n.d.). https://blog.americansafetycouncil.com/the-history-of-speed-limits-in-america/

Ashford, E., Mulgan, T.: Contractualism [Stanford Encyclopedia of Philosophy]. Stanford Encyclopedia of Philosophy, 30 August 2007. https://plato.stanford.edu/entries/contractualism/#WhaCon. Accessed 27 May 2024

Automated Vehicle Safety Consortium: AVSC Best Practice for Metrics and Methods for Assessing Safety Performance of Automated Driving Systems [ADS]. SAE Industry Technologies Consortium (2021). https://www.sae.org/standards/content/avsc00006202103/

Automated Vehicle Safety Consortium: AVSC Best Practice for Core Automated Vehicle Safety Information. SAE Industry Technologies Consortia (2024). https://www.sae.org/standards/content/avsc-d-02-2024/

AZ DPS: Help Prevent Secondary Collisions by Moving Minor Incidents Off of the Highway. Arizona Department of Public Safety (2022). https://www.azdps.gov/news/digest/112

Bärgman, J., Lisovskaja, L., Victor, T., Flannagan, C., Dozza, M.: How does glance behavior influence crash and injury risk? A 'what -if' counterfactual simulation using crashes and near-crashes from SHRP2. Transp. Res. Part F **35**, 152–169 (2015)

Bartlett, J.S.: Cars, SUVs, and Trucks With the Best and Worst Braking Distances. Consumer Reports, 12 January 2021. https://www.consumerreports.org/cars/car-safety/best-and-worst-braking-distances-a2960086475/

Bicchieri, C.: The Rules We Live By. The Nature and Dynamics of Social Norms. Cambridge University Press, In: The Grammar of Society (2006). https://doi.org/10.1093/acprof:oso/9780190622046.001.0001

Bicchieri, C., Demo, L.G.: Norm strength and norm stability. Curr. Opin. Psychol. **62**,(2024). https://doi.org/10.1016/j.copsyc.2024.101957

Bicchieri, C., Muldoon, R., Sontuoso, A.: Social Norms. Stanford Encyclopedia of Philosophy, 1 March 2023. https://plato.stanford.edu/entries/social-norms/. Accessed 6 Nov 2024

Bin-Nun, A.Y., Derler, P., Mehdipour, N., Duintjer Tebbens, R.: How should autonomous vehicles drive? Policy, methodological, and social considerations for designing a driver. Hum. Soc. Sci. Commun. **9** (2022)

Bonnefon, J.-F., Rahwan, I., Shariff, A.: The moral psychology of artificial intelligence. Annu. Rev. Psychol. **75** (2024)

Bouhsissin, S., Sael, N., Benabbou, F.: Driver behavior classification: a systematic literature review. IEEE Access **11** (2023). https://doi.org/10.1109/ACCESS.2023.3243865

Boyle, L.L.: The 2021 National Survey of the Use of Booster Seats. CrashStats - NHTSA, March 2023. https://crashstats.nhtsa.dot.gov/Api/Public/ViewPublication/813396.pdf. Accessed 6 Nov 2024

British Standards Institution: Digital commentary driving: the new safety technique. BSI Group, 16 June 2021. https://www.bsigroup.com/en-GB/insights-and-media/media-centre/press-releases/2021/june/digital-commentary-driving-the-new-safety-technique-that-can-help-put-automated-vehicles-on-our-roads/

CA DMV: Article 3.7 Testing of Autonomous Vehicles (2022). https://www.dmv.ca.gov/portal/file/adopted-regulatory-text-pdf/

CDC: Increased Safety-Belt Use – United States, 1991. MMWR **41**(24), 421–423 (1992)

Chaloux, N., Ahmed, A.: Rules of the road: compliance and defiance among the different types of cyclists. Transp. Res. Rec. **2673**(9), 34–43 (2019)

D'Amato, A., Dancel, S., Pilutti, J., Tellis, L., Frascaroli, E., Gerdes, J.C.:. Exceptional driving principles for autonomous vehicles. J. Law Mob. (2022)

Defense Science Board: The Role of Autonomy in DoD Systems. Department of Defense (2012)

De Freitas, J., Censi, A., Walker Smith, B., Di Lillo, L., Anthony, S.E., Frazzoli, E.: From driverless dilemmas to more practical commonsense tests for automated vehicles. Proc. Natl. Acad. Sci. USA **118**(11) (2021)

De Freitas, J., Zhou, X., Atzei, M., Boardman, S., Di Lillo, L.: Consumers Hold Autonomous Vehicles Liable Even When Not at Fault. Working Paper 23-036 (2023)

Dinparastdjadid, A., Supeene, I., Engstrom, J.: Measuring surprise in the wild. arXiv https://arxiv.org/pdf/2305.07733 (2023)

Doubek, F., Salzmann, F., de Winter, J.: What makes a good driver on public roads and race tracks? An interview study. Transp. Res. Part F **80**, 399–423 (2021)

Dushyant: Ford Mustang Dark Horse Breaks Records with Shortest Stopping Distance. WheelCircuit, 6 October 2023. https://www.wheelcircuit.com/blog/ford-mustang-dark-horse-breaks-records-with-shortest-stopping-distance

ECE: Addendum 156 – UN Regulation No. 157 Amendment 4. Concerning the Adoption of Harmonized Technical United Nations Regulations for Wheeled Vehicles, Equipment and Parts which can be Fitted and/or be Used on Wheeled Vehicles and the Conditions for Reciproc... UNECE, 3 March 2023 https://unece.org/transport/documents/2023/03/standards/un-regulation-no-157-amend4. Accessed 5 Sept 2024

Ecola, L., Popper, S.W., Silberglitt, R., Fraade-Blanar, L.: The Road to Zero. RAND Corporation (2018)

Engstrom, J., Christophe, E., Lee, J., Supeene, I., Mathias, R.:. US Patent 11,834,070 (2023)

Engstrom, J., Liu, S.-Y., Dinparastdjadid, A., Simoiu, C.: Modeling road user response timing in naturalistic traffic conflicts: a surprise-based framework. Acc. Anal. Prevent. **198** (2024)

Eschner, K.: Henry Bliss, America's First Pedestrian Fatality, Was Hit By an Electric Taxi. Smithsonian Magazine, 13 September 2017. https://www.smithsonianmag.com/smart-news/henry-bliss-americas-first-pedestrian-fatality-was-hit-electric-taxi-180964852/. Accessed 24 May 2024

Fabio, U.D., et al.: *Ethics Commission: Automated and Connected Driving*. Federal Ministry of Transport and Digital Infrastructure [Germany] (2017). https://bmdv.bund.de/SharedDocs/EN/publications/report-ethics-commission.pdf?__blob=publicationFile

Fatally Hurt by Automobile. New York Times, 14 September 1899. https://timesmachine.nytimes.com/timesmachine/1899/09/14/102413700.pdf?pdf_redirect=true&ip=0

Favarò, F., et al.: Building a Credible Case for Safety: Waymo's Approach for the Determination of Absence of Unreasonable Risk. Waymo (2023)

Favaro, F. M., Agarwal, S., Nader, N., Mahmood, S.: Towards a Smart World: Hazard Levels for Monitoring of Autonomous Vehicles' Swarms. SJSU (2018). https://transweb.sjsu.edu/sites/default/files/1735_Favaro-Autonomous-Vehicle-Swarms.pdf

Fraade-Blanar, L., Blumenthal, M.S., Anderson, J.M., Kalra, N.: Measuring Automated Vehicle Safety: Forging a Framework. RAND Corporation (2018)

Freeman, S.: Original Position [Stanford Encyclopedia of Philosophy]. Stanford Encyclopedia of Philosophy, 20 December 2008. https://plato.stanford.edu/entries/original-position/#VeilIgno. Accessed 29 May 2024

Friedman, B., College, C., Institute, M.: Value-Sensitive Design. Interactions, November - December (1996). https://dl.acm.org/doi/pdf/10.1145/242485.242493

Friedman, B., Kahn, P.H., Borning, A., Zhang, P.: Value sensitive design and information systems. In: Zhang, P., Galletta, D.F. (eds.) Human-Computer Interaction and Management Information Systems: Foundations. Taylor & Francis Group (2015)

Geisslinger, M., Poszler, F., Betz, J., Lütge, C., Lienkamp, M.: Autonomous driving ethics: from trolley problem to ethics of risk. Philos. Technol. **34**, 1033–1055 (2021)

Gerdes, J.C.: The virtues of automated vehicle safety-mapping vehicle safety approaches to their underlying ethical frameworks. In: 2020 IEEE Intelligent Vehicles Symposium [IV], 107–113 (2020)

Hang, P., Chen, X.: Towards autonomous driving: review and perspectives on configuration and control of four-wheel independent drive/steering electric vehicles. Actuat. Intell. Electr. Vehic. **10**(8) (2021)

Himma, K.E., Tavani, H.T. (eds.): The Handbook of Information and Computer Ethics. Wiley (2008)

Hulse, L., Xie, H., Galea, E.R.: Perceptions of autonomous vehicles: Relationships with road users, risk, gender and age. Saf. Sci. **102** (2018)

IEEE: 7000-2021 - IEEE Standard Model Process for Addressing Ethical Concerns during System Design (2021). https://ieeexplore.ieee.org/document/9536679

IEEE: IEEE 2846-2022: Standard for Assumptions in Safety-Related Models for Automated Driving Systems (2022). https://standards.ieee.org/ieee/2846/10831/

IEEE: IEEE 3321–2024: IEEE Approved Draft Recommended Practice for the Application of Assumptions on Reasonably Foreseeable Behavior of Other Road Users in Safety Related Models (2024). https://standards.ieee.org/ieee/3321/11616/

IIHS: Alcohol and drugs. IIHS (2024a). https://www.iihs.org/topics/alcohol-and-drugs. Accessed 16 Oct 2024

IIHS: *Speed/Speed limits*. IIHS (2024b). https://www.iihs.org/topics/speed. Accessed 16 Oct 2024

ISO: ISO 26262-1:2018 Road vehicles — Functional safety (2018). https://www.iso.org/standard/68383.html

ISO: ISO 21448:2022 Road vehicles — Safety of the intended functionality (2022). https://www.iso.org/standard/77490.html

ISO: ISO 39003:2023 Road traffic safety [RTS] — Guidance on ethical considerations relating to safety for autonomous vehicles (2023). https://www.iso.org/standard/50665.html
Itkonen, T.H., Pekkanen, J., Lappi, O., Kosonen, I., Luttinen, T., Summala, H.: Trade-off between jerk and time headway as an indicator of driving style. *PLoS ONE* **12**(10) (2017)
Japan Automobile Manufacturers Association, Inc.: Automated Driving Safety Evaluation Framework Ver 3.0. Sectional Committee of AD Safety Evaluation, Automated Driving Subcommittee (2022). https://www.jama.or.jp/english/reports/docs/Automated_Driving_Safety_Evaluation_Framework_Ver3.0.pdf
Kongsvik, T., Haavik, T., Bye, R., Almklov, P.: Re-boxing seamanship: From individual to systemic capabilities. Saf. Sci. **130**(Oct) (2020)
Lee, J.D., See, K.A.: Trust in automation: designing for appropriate reliance. Hum. Fact. **46**(1), 50–80 (2004)
Lee, T., Son, J.: Relationships between driving style and fuel consumption in highway driving. SAE Technical Paper (2011)
Lie, A., Tingvall, C.: Are crash causation studies the best way to understand system failures – Who can we blame? Acc. Anal. Prevent. **196**(Mar) (2024)
Mattas, K., et al.: Driver models for the definition of safety requirements of automated vehicles in international regulations. Application to motorway driving conditions. Acc. Anal. Prevent. **174**(Sept.) (2022)
Mattas, K., et al.: Fuzzy Surrogate Safety Metrics for real-time assessment of rear-end collision risk. A study based on empirical observations. Acc. Anal. Prevent. **148**(Dec) (2020)
McKnight, A.J., Adams, B.B.: Driver education task analysis: task analysis methods. HumRRO Technical Report 72-13 (1972)
Michon, J.A.: A critical view of driver behavior models: what do we know, what should we do? In: Evans, L., Schwing, R.C. (eds.) Human Behavior and Traffic Safety, pp. 485–520. Springer US (1985). https://doi.org/10.1007/978-1-4613-2173-6_19
Miller, K.: Designing Ethical Self-Driving Cars. Stanford HAI, 23 January 2023. https://hai.stanford.edu/news/designing-ethical-self-driving-cars. Accessed 28 Aug 2024
Nair, G.S., Bhat, C.R.: Sharing the road with autonomous vehicles: Perceived safety and regulatory preferences. Transp. Res. Part C: Emerg. Technol. **122**(Jan) (2021)
NCSL: State Cellphone Use While Driving Laws. National Conference of State Legislatures, 30 September 2024. https://www.ncsl.org/transportation/distracted-driving-cellphone-use. Accessed 20 Oct 2024
Nergard, V.: Airmanship – a qualitative approach. Aviation **18**(3) (2014)
Nichols, J.L., Solomon, M.G.: Click It or Ticket Evaluation, 2011 [DOT HS 811 779 ed.]. National Highway Traffic Safety Administration (2013)
Nister, D., Lee, H.-L., Ng, J., Wang, Y.: The Safety Force Field. Nvidia, 1 March 2019. https://www.nvidia.com/content/dam/en-zz/Solutions/self-driving-cars/safety-force-field/the-safety-force-field.pdf. Accessed 7 Oct 2024
Nordfjærn, T., Şimşekoğlu, Ö.: Empathy, conformity, and cultural factors related to aberrant driving behaviour in a sample of Urban Turkish drivers. Saf. Sci. **68**, 55–64 (2014)
NYPD: History of Traffic. New York City Policy Department, CWA Local 1182 (n.d.). https://local1182.org/about-us/history-of-traffic/
Oh, S., Avedisov, S., Orosz, G.: On the handling of automated vehicles: Modeling, bifurcation analysis, and experiments. Eur. J. Mech. **90**(Nov-Dec)
Oliveira, L., Proctor, K., Burns, C., Birrell, S.: Driving style: how should an automated vehicle behave? Information **10**(6) (2019)
Olleja, P., Markkula, G., Bargman, J.: Validation of human benchmark models for Automated Driving System approval: How competent and careful are they really? (2024). *arXiv*. https://arxiv.org/abs/2406.09493

Othman, K.: Public acceptance and perception of autonomous vehicles: a comprehensive review. AI Ethics **1**, 355–387 (2021)

Parker, D., Reason, J. T., Manstead, A.S., Stradling, S.: Driving errors, driving violations and accident involvement. Ergonomics **38**(5), 1036–1048 (2007)

Pecheux, K.K., Carrick, G., Pecheux, B.: Secondary crash research: a multistate analysis. FHWA-HOP-23-043 (2023). https://ops.fhwa.dot.gov/publications/fhwahop23043/fhwahop23043.pdf

Peiris, A., Harik, I.E.: Vehicle on Shoulder and Crash — Correlation or Causation? University of Kentucky Transportation Center (2019). https://uknowledge.uky.edu/ktc_researchreports/1624/

Pek, C., Manzinger, S., Kochii, M., Althoff, M.: Using online verification to prevent autonomous vehicles from causing accidents. Nat. Mach. Intell. **2**, 518–528 (2020)

Peng, C., et al.: Drivers' evaluation of different automated driving styles: is it both comfortable and natural? Hum. Fact. **66**(3) (2024)

Peng, H., Zhu, M., Lenneman, J.K.: Automated Vehicles and Roadmanship: How Safe Robot Cars Can Read the Road Like Humans Do. Mcity, 23 February 2022. https://mcity.umich.edu/wp-content/uploads/2022/03/Automated-Vehicles-and-Roadmanship.pdf. Accessed 5 Sept 2024

Pyrialakou, V.D., Gkartzonikas, C., Gatlin, J.D., Gkritza, K.: Perceptions of safety on a shared road: driving, cycling, or walking near an autonomous vehicle. J. Safety Res. **72**, 249–258 (2020)

Reason, J.: Errors and violations on the roads: a real distinction? Ergonomics **33**(10/11), 1315–1332 (1990)

Reddy, K., Nassif, E., Angeloudis, P., Quddus, M.: Towards a Universal Evaluation Model for Careful and Competent Autonomous Driving (2024). arXiv https://doi.org/10.48550/arXiv.2407.15596

Roy, M.M., Liersch, M.J.: I am a better driver than you think: examining self-enhancement for driving ability. J. Appl. Soc. Psychol. **43**(8), 1648–1659 (2013)

SAE: J3016_202104: Taxonomy and Definitions for Terms Related to Driving Automation Systems for On-Road Motor Vehicles. SAE International (2021). https://www.sae.org/standards/content/j3016_202104/

SAE: J3164_202301: Ontology and Lexicon for Automated Driving System [ADS]-Operated Vehicle Behaviors and Maneuvers in Routine/Normal Operating Scenarios. SAE International (2023). https://www.sae.org/standards/content/j3164_202301/

SAE: J3330 recommended practice on behavior reference models for evaluating automated driving systems in traffic conflict scenarios. SAE International (2024). https://www.sae.org/standards/content/j3330/

Scanlon, J.M., Kusano, K.D., Engström, J., Victor, T.: Collision avoidance effectiveness of an automated driving system using a human driver behavior reference model in reconstructed fatal collisions. Waymo.com (2022). https://waymo.com/research/collision-avoidance-effectiveness-of-an-automated/. Accessed 6 Sept 2024

Scanlon, T.M.: What We Owe to Each Other. Harvard University Press (1998)

Scanlon, T.M.: Contractualism and utilitarianism. In: Sen, A., Williams, B. (eds.) Utilitarianism and Beyond. Cambridge University Press (2010)

Schwarting, W., Pierson, A., Mora, J.A., Karaman, S.: Social behavior for autonomous vehicles. PNAS **116**(50) (2019)

Shalev-Shwartz, S., Shammah, S., Shashua, A.: On a Formal Model of Safe and Scalable Self-driving Cars, 27 October 2018. arXiv. https://arxiv.org/pdf/1708.06374

Solomon, M.G., Compton, R.P., Preusser, D.F.: Taking the click it or ticket model nationwide. J. Saf. Res. **35**(2) (2004)

Tennant, C., Neels, C., Parkhurst, G., Jones, P., Mirza, S., Stilgoe, J.: Code, culture, and concrete: self-driving vehicles and the rules of the road. Front. Sust. Cit. **3**(710478) (n.d.)

Texas Transportation Code: Texas Transportation Code 550.022 (2005). https://statutes.capitol.texas.gov/Docs/TN/pdf/TN.550.pdf

Tillmann, W.A., Hobbs, G.E.: The Accident-prone automobile Driver: a study of the psychiatric and social background. Am. J. Psychiatry **106**(5) (1949)

Tingvall, C., Lie, A.: The concept of "acceptable risk" applied to road safety risk level. In: Vickerman, R. (ed.) International Encyclopedia of Transportation, pp. 2–5. Elsevier (2021)

Townsend, B., et al.: From pluralistic normative principles to autonomous-agent rules. Mind. Mach. **1**(4), 1–33 (2022)

UL 4600: ANSI/UL 4600 Standard for Safety for the Evaluation of Autonomous Products. UL Standards & Engagement (2023). https://ulse.org/ul-standards-engagement/autonomous-vehicle-technology. Accessed 16 Oct 2024

US DOT: The Haddon Matrix. Highway Safety Improvement Program Manual – Safety—Federal Highway Administration (2011). https://safety.fhwa.dot.gov/hsip/resources/fhwasa09029/app_c.cfm. Accessed 16 Oct 2024

Uys, P.E., Els, P.S., Thoresson, M.J.: Criteria for handling measurement. J. Terrramech. **43**(1), 43–67 (2006)

Uzgalis, W.: John Locke [Stanford Encyclopedia of Philosophy]. Stanford Encyclopedia of Philosophy, 2 September 2001. https://plato.stanford.edu/entries/locke/. Accessed 29 May 2024

Vinkhuyzen, E., Cefkin, M.: Developing socially acceptable autonomous vehicles. EPIC 2016, **2016**(1), November 2016

Wang, J., Sun, F., Ge, H.: Effect of the driver's desire for smooth driving on the car-following model. Physica A **512**, 96–108 (2018)

Wang, Y., Ren, Y., Elliott, S., Zhang, W.: Enabling courteous vehicle interactions through game-based and dynamics-aware intent inference. IEEE Trans. Intell. Vehic. **5**(2) (2019)

Waymo: Drivership. Facebook, 19 July 2022, https://www.facebook.com/Waymo/videos/drivership/1498304117289295/. Accessed 27 May 2024

Webb, N., et al.: Waymo's safety methodologies and safety readiness determinations (2020). arXiv https://arxiv.org/abs/2011.00054

Weng, B., Rao, S. J., Deosthale, E., Schnelle, S., Barickman, F.: Model predictive instantaneous safety metric for evaluation of automated driving systems. In: 2020 IEEE Intelligent Vehicles Symposium (2020)

Wijayasekara, D., Manic, M., Gertman, D.: Driving behavior prompting framework for improving fuel efficiency. In: 2014 7th International Conference on Human System Interactions [HSI], pp. 55–60 (2014)

Williams, K.: Who Was the First Person Killed by a Car in New York? [Published 2018]. The New York Times, 30 March 2018. https://www.nytimes.com/2018/03/30/nyregion/who-was-the-first-person-killed-by-a-car-in-new-york.html. Accessed 27 May 2024

Xiao, W., et al.: Rule-based Optimal Control for Autonomous Driving (2021). arXiv https://arxiv.org/abs/2101.05709

Yuen, K.F., Chua, G., Wang, X., Li, K.X.: Understanding public acceptance of autonomous vehicles using the theory of planned behaviour. Int. J. Environ. Res. Pub. Health **17**(12) (2020)

Towards a Unified Process Model for User-Centered Cooperative Intelligent Transportation System Research and Development

Fabian Hub[1(✉)], Marc Wilbrink[2], Tobias Hesse[1], and Michael Oehl[2]

[1] Institute of Transportation Systems, German Aerospace Center (DLR), Rutherfordstraße 2, 12489 Berlin, Germany
{fabian.hub,tobias.hesse}@dlr.de
[2] Institute of Transportation Systems, German Aerospace Center (DLR), Lilienthalplatz 7, 38108 Braunschweig, Germany
{marc.wilbrink,michael.oehl}@dlr.de

Abstract. Research projects in the field of Cooperative Intelligent Transport Systems (C-ITS) are characterized by high complexity. Coordinating various mutually dependent subsystems throughout the development process, while simultaneously ensuring responsiveness and quality assurance, represents a key challenge for future large-scale projects. Since existing process models often fail to manage such complexity in practice, the need for new development approaches in C-ITS has become evident. Building on four established process models, we combined their respective advantages and designed a new, unified, and user-centered process model for multi-layered C-ITS projects. Results from a subsequent evaluation workshop suggest that this unified and user-centered model can enhance cooperation and reduce complexity for practitioners. By proposing an alternative approach to structuring C-ITS projects, this work contributes to advancing sustainable, highly connected transportation and mobility systems and addresses current challenges in the sector more effectively.

Keywords: cooperative intelligent transport system · cooperative connected and automated mobility · process model · user-centered · iterative · expert workshop · research & development · human cyber physical system · system-of-systems

1 Introduction

The challenge for research in the field of Cooperative Intelligent Transport Systems (C-ITS) is to develop innovative, connected solutions for the successful transformation of mobility and transport. To achieve this, it is essential to explore and incorporate the potential of all interconnected systems and perspectives. Hence, this work proposes a novel approach for research and development (R&D) processes in C-ITS. Focusing on collaboration, user-centeredness, and iteration, the proposed process model is intended to enhance the effectiveness and efficiency of deeply integrated and highly complex research projects.

G. Meyer and S. Beiker (Eds.): ARTSymposium 2024, LNMOB, pp. 106–123, 2026.
https://doi.org/10.1007/978-3-032-12374-9_7

The interdependencies of C-ITS systems present unique challenges due to the diversity and interconnectedness of subsystems, including humans, automated vehicles, connectivity, traffic infrastructure, and overall traffic management (Ranjbar Motlagh et al. 2024). Although traditional process models such as the widely adopted V-Model (Rook 1986; VDI/VDE 2206) provide robust structures for technical development and project management, they become limited in addressing highly integrated R&D projects. Process models remain essential tools for managing complexity in system development: by visualizing workflows and identifying dependencies across project activities, they guide practitioners and help ensure high-quality outputs.

However, addressing multiple complex subsystems, as required in C-ITS development, demands process models that can manage dynamic changes at the subsystem level with integrated feedback loops and continuous interdependent reflection within workflows. For example, the European Commission's initiative on connected, cooperative, and automated mobility (CCAM) responds to these challenges with conventional models such as the Festa V (FOT-Net & CARTRE 2018). Yet, because such models establish requirements primarily at the outset, their linear and rigid structures often fail to adequately reflect the dynamic interactions that are critical in large-scale mobility projects.

At the advent of widespread CCAM/C-ITS deployment, research projects at local, national, and international levels aim to deliver successful demonstrations at scale (30 European sites by 2030), as well as pilots and research integrated with commercial operations (CCAM Strategic Research and Innovation Agenda, European Commission 2023). Current hurdles include the limited scalability and operationalization of user-oriented, integrated CCAM/C-ITS solutions for passenger and freight mobility. To overcome these hurdles, methodologies are needed to validate the safe system functions of CCAM use cases and to establish common evaluation frameworks that support the exchange and reuse of R&D results. In this context, C-ITS projects require development processes that enable modularization to handle rising system complexity, ensure multi-system perspective design and evaluation for real-world applicability, and provide measurable impacts on mobility and transport systems. Iteration is equally essential to ensure adaptability to changing conditions and to address unpredictable consequences, while maintaining compatibility with real-world operations. This highlights the need for more adaptable, modular, iterative, and holistic process models in C-ITS that ensure resource efficiency, broaden the knowledge base, and promote sustainable, safe, user-centered, and inclusive transportation systems. Furthermore, developing processes that streamline collaboration across subsystems, human factors, automated vehicles, connectivity, infrastructure, and traffic, while fostering continuous learning is key to the success of future CCAM and C-ITS projects.

The coordination of mutually dependent subsystems (human–automation–infrastructure–connectivity–traffic) throughout the development process, while maintaining openness of research and ensuring sustainable quality assurance, remains a central challenge. Consequently, a restructured approach to research and development in future C-ITS projects is required. Thus, this work poses the following question: *What should a process model for C-ITS research projects look like to reduce complexity in multi-system development, respond appropriately to changes during R&D activities, and ensure continuous and sustainable improvement of the target system?*

Our proposed unified process model for iterative and user-centered innovation and R&D represents a novel approach to C-ITS development processes by integrating the strengths of well-established design models into the C-ITS context. Specifically, we combine the V-Model's (Rook 1986) emphasis on validation and verification, the DevOps model's (Debois 2008) focus on continuous integration and delivery, the Spiral Model's (Boehm 1988) iterative perspective and strategic research orientation, and the User-Centered Design process (ISO 9241-210:2019) with its iterative structure and focus on user needs, into one unified framework.

This unified model addresses the challenges of C-ITS by supporting complexity management and fostering collaboration and efficiency in R&D through a combination of iterative steps and dedicated verification and validation phases for all subsystems involved. In doing so, it contributes to making research and innovation more sustainable, especially with regard to technology transfer and dissemination.

To validate its effectiveness and impact, this work presents both the development and evaluation of the new process model, specifically designed to improve complex C-ITS systems. R&D projects in this domain must advance efficiently and effectively, accounting for the many subsystems and their interdependencies. To receive initial feedback about the proposed model, we conducted an online workshop with four practitioners experienced in successful C-ITS research projects. The evaluation focused on applicability, potential benefits, challenges, and opportunities for improvement from their perspective. As a result, this work provides a novel perspective on managing complex and multi-faceted C-ITS projects by highlighting the user-centered perspective and integrating the strengths of well-established design models to improve collaboration, research effectiveness, and practical contributions to future mobility and transportation systems.

2 Methodological Approach

This work is structured in the following sections: First, an overview of existing and widely adopted process models for design and development and their contribution to a new and unified process model for C-ITS projects its outlined. Thereafter, the new unified user-centered C-ITS process model is presented. Third, this work presents the qualitative evaluation of the new process model conducted by practitioners in the field. Finally, concluding results are discussed and future improvements are presented.

2.1 Overview of the Used Process Models

The next section presents four commonly used process and design models, which served as basis for the development of a new process model specifically for the field of C-ITS.

2.1.1 V-Model

The V-Model, is an extension of the Waterfall Model (Royce 1987) and widely applied across various industries, like the Festa V for field operational tests in C-ITS projects

(FOT-Net & CARTRE, 2018; Jeong and Heo 2020; Khudeynatov and Hashimov, 2023; Lim and Chin, 2020; Lu et al. 2019). It is characterized by its eponymous "V" shape (see Fig. 1), which organizes the development process into phases of defining requirements, development and system implementation, and integration and testing (Lim & Chin, 2020), each comprising a set of activities. Usually, various subsystems, like software, electronics, mechanics and others are brought together to parallelly follow the same structure (VDI/VDE 2021). Therefore, the R&D plans and activities can be tracked easily. The requirements phase decomposes the system development downstream into subtasks (Graessler 2017; Graessler and Hentze, 2020; Mathur and Malik, 2010; Lim and Chin 2020). During development and implementation at subsystem level, functional testing is included. In contrast, the integration and testing phase focuses on the merging all components upstream into the system. Ultimately, verification ensures that the system is engineered efficiently, validation ensures that the system effectively meets requirements (Graessler 2017; Lake 1999). Despite wide-spread industry application, the V-Model is not without shortcomings. Although the large amount of documentation ensures structure, quality and compliance (Deuter 2013), time- and cost-efficiency is reduced (Amelio 2010; Pressman 2014). Since requirements and test documentation are specified early on in the development process, it lacks flexibility (Regulwar et al. 2010; Li 2015). Another issue related to the inflexibility is the sequential nature of the model, posing a challenge to adapt along the process, for example to implement findings along the way (George 2023). Concluding, the emphasis on consequent validation and verification feedback is beneficial for large and complex research projects. Also, the inherent consideration of various subsystems makes it viable. Hence, those two aspects should be represented in the new C-ITS process model.

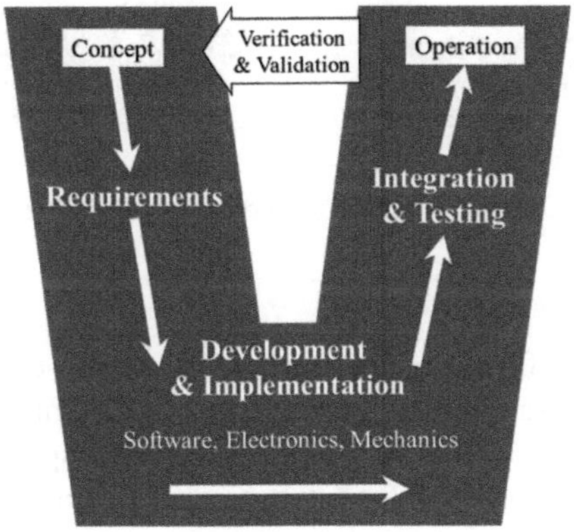

Fig. 1. Schematic V-Model, own representation based on VDI/VDE (2021).

2.1.2 DevOps

The concept of DevOps (Debois 2008), puts a strong emphasis on fostering collaboration (Leite et al. 2019) between traditionally siloed development (Dev) and operations teams (Ops) (Humble and Farley 2010; Leite et al. 2019) (see Fig. 2). Its primary goal is to promote faster software delivery and quicker user/customer feedback (Humble and Farley 2010; Lwakatare et al. 2016).

DevOps' core strength lies in its adaptability to rapid changing project requirements and fast error detection as well as update releases (Debois 2008), thereby it found wide application in software development (Erich et al. 2017; Rembetsy and McDonell 2012; Neely and Stolt, 2013; Balalaie et al. 2016; Bass et al. 2015).

Yet, DevOps is not without challenges. Failing to meet the high cooperation and communication requirements can lead to uncoordinated activities, resulting in inefficiencies and suboptimal outcomes (Iden et al. 2011; Tessem and Iden, 2008; Riungu-Kalliosaari et al. 2016). The complexity of integrating numerous processes, tools, and technologies within the DevOps framework can also overwhelm practitioners (Humble and Farley 2010), particularly in organizations with rigid, deeply ingrained cultures resistant to adopting agile methods central to DevOps (Riungu-Kalliosaari et al. 2016; Poppendieck and Poppendieck, 2003).

The DevOps' focus on continuous integration and delivery is a strength that could emphasize feedback loops between C-ITS subsystems and should therefore be reflected in a new C-ITS project framework.

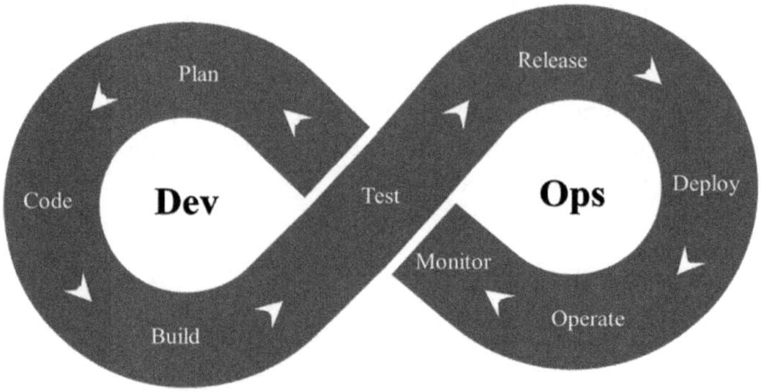

Fig. 2. Schematic DevOps cycle model, own representation based on Atlassian (2025).

2.1.3 User Centered Design

User-Centered Design (UCD) is rooted in the fields of human-computer interaction (HCI) (Norman and Draper 1986) and has influenced a growing number of methodologies (Norman 2013; Brown 2008; Brown 2009). At its core, UCD emphasizes understanding user needs, preferences and behaviors before starting engineering activities (Norman 2013). Essentially, instead of relying on assumptions about system requirements, UCD places the user's actual problems at the forefront of design decisions (see Fig. 3). According to ISO 9241–210, UCD is "an approach to systems design and development that aims to make interactive systems more usable by focusing on the use of the system and applying human factors/ergonomics and usability knowledge and techniques" (International Organization for Standardization, 2010). UCD consists of several activities performed in an iterative manner, comprised by understanding of context and environment, specifying requirements, designing prototypical solutions and evaluating these by utilizing user feedback. It emphasized a continuous and iterative refinement until arriving at a solution that is both functional and desirable for users (Sharp et al. 2019; Norman 2013; George 2008).

Although UCD is broadly accepted to ensure effective output, its iterative nature, stressing continuous investigating, building and testing, can be resource intensive (Norman 2013; Sharp, Rogers et al. 2019).

Concluding, the iterative, yet structured approach of UCD to finding solutions for diligently investigated user needs should be incorporated into the new C-ITS process model, since it is of upmost importance to address the human perspective in future mobility and transportation systems.

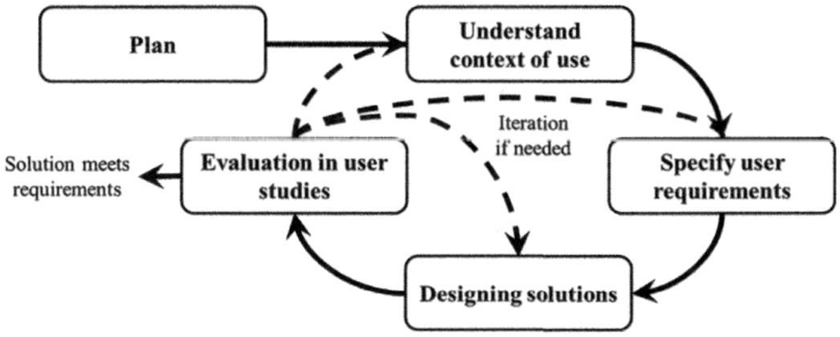

Fig. 3. Schematic user-centered design process, own representation based on ISO 9241-210.

2.1.4 Spiral Model

The Spiral Model (Boehm 1988), is a flexible meta-model that integrates elements from the Waterfall Model (Royce 1987), iterative cyclical processes, and prototyping-oriented models (Pomberger 2006). Its adaptability has led to widespread application across various domains, including public projects like aerospace and defense (Viravan 1997; Boehm and Hansen 2001; Pressman 2014).

Characterized by its iterative nature, the Spiral Model employs a dual focus on cumulative costs (radial dimension) and progress through iterations (angular dimension) (Ruparelia 2010; Boehm 1988). Each "spiral" cycle consists of four key steps: planning, analysis, engineering and evaluation (see Fig. 4). These steps allow for systematic identification and mitigation of risks early in the development process, supported by continuous feedback from stakeholders (Boehm 1988; Sommerville, 2010). The model's iterative structure is particularly effective in high-risk contexts, where its emphasis on risk management helps reduce unforeseen costs and misaligned requirements (Munassar and Govardhan 2010; Mujumdar et al. 2012).

Despite its strengths, especially compared earlier, more linear models (Boehm 1988; Nilsson and Wilson 2012; Nuseibeh 2001), its focus on risk identification and remediation requires specialized expertise (Mujumdar et al. 2012). Additionally, it does not prioritize the sustainable use of existing components, a key aspect in today's R&D practices (Pomberger 2006).

Overall, the Spiral Model remains an advanced strategic approach in R&D, offering risk assessment while accommodating dynamic project requirements (Boehm 1988;

Sommerville 2010). Consequently, its iterative perspective and focus on long-term R&D trajectories should be a reflected in the new C-ITS process model.

Fig. 4. Schematic spiral model, own representation based on Boehm (1988) and Pomberger (2006).

2.2 Developing the New Unified User-Centered C-ITS Process Model

This upcoming paragraph describes the new unified C-ITS process model and its development. Subsequently, the evaluation phase is shown.

In the field of C-ITS, project complexity is increasing with further system integration and therefore needs an adequate R&D framework, which is currently not existing. Hence, goal was to construct a framework for design and development in the research field of multi system C-ITS projects.

2.2.1 Process Model Unification and Development

The authors, with backgrounds in psychology and engineering, conceptualized the new and unified C-ITS process model in about one dozen online workshop sessions, spanning approximately 2h each (preparation and debriefing excluded). The approach was to take the strengths of the prominent R&D frameworks and to blend them into one unified and user-centered process model that addresses challenges and reduces complexities of future, highly integrated projects that aim to improve the state of future transportation and mobility. As a starting point, the well-known and in C-ITS widely acknowledged Festa V (FOT-Net & CARTRE, 2018) for so-called Field Operational Tests, and it's three-step structure (preparing, using and evaluating) was utilized to maintain a high level of transferability and familiarity. Additionally, a former approach to restructuring C-ITS projects was reviewed (Schindler and Hesse 2016). Also, aspects of the research context that influence goals, activities and output were transferred and further specified. As all investigated process models show successive progress, the new unified process model also presents the steps in a stage-gate structure. In the next steps, aspects for iteration were implemented. Hence, applying a round shape structure and a clock-wise procedure for the model visualization with potential diverging decisions and the ever-evolving characteristics of R&D, just like the UCD process and the Spiral Model, seemed favorable. In a third step, the different subsystems of a C-ITS project were addressed by adding multiple layers in the center model. The subsystems go from a very specific human/user-centered system at the core, across various technical C-ITS-components (automation, connectivity, infrastructure) to a broader traffic system at the shell. The user perspective is placed in the center to emphasize the importance of the human perspective in C-ITS. In that regard, we advocate for establishing user-centered problem statements as a baseline for starting the R&D activities. It is important to note that the model does not define all interdependencies accurately across these subsystems, but rather tries to reflect each perspective and respect general interactions in the R&D process.

2.2.2 The New Unified User-Centered Process Model for C-ITS Projects

The new model represents the systematic and integrated design, development, operational using and evaluation of Cooperative, Connected, and Automated Mobility (C-ITS) systems with the aim of continuous and sustainable system improvement. The aim of the model is the transition from an existing C-ITS system (t_n) to an improved version of a future state (t_{n+1}) through a cycle of iterative activities (see Fig. 1).

As the overall goal of C-ITS projects is to improve the status quo in mobility systems with highly connected automation, infrastructure, communication and traffic components it is important to acknowledge the existing context and the challenges in the initial starting point for R&D of the respective mobility system.

Since the R&D activities cannot change the context directly, it influences the status quo and all activities. Here, the context is broadly described by mega trends, local challenges, governmental and legal frameworks to act within, but also specific goals/paradigms of today's time, like climate protection, managing transformation, and ensuring mobility. Since these contextual factors heavily influence the activities in C-ITS, they should be addressed in the overall project goals. The new unified user-centered C-ITS process model tries to improve C-ITS in the long run and ultimately change the context, too.

At the core of the new unified and user-centered process lies the integrated C-ITS system-of-systems, which consists of interrelated subsystems and perspectives: human, automation, connectivity (e.g., vehicle-to-everything; V2X), infrastructure, and traffic, represented in an onion format (Saunders et al., 2016). This subsystem structure could also be referred to as a system of human-cyber-physical systems (HCPS). While development for each of the subsystems holds approaches of its own, bringing these subsystems together to an integrated and effective system by improving cooperation, reliability, quality and efficiency throughout the interdisciplinary project requires this new form of organizing the R&D activities.

Overall, the process structures the activities into three key phases (see Fig. 5): *Preparing* (fat arrow on left side, starting with "problem Δ solution"), *Using* (fat arrow on the bottom, starting with "integrated functional prototype"), and *Analyzing* (fat arrow on right side, starting with "collected raw data"). A distinctive stage-gate approach characterizes each phase, ensuring structured progression while allowing for iterative refinement both within individual subsystems and in the HCPS as a whole. It is important to note that the process is to be seen as modular and can be adapted according to the project's scope, complexity, and expectations. For instance, smaller projects do not include all the presented HCPS subsystems, but the general process outlined should be applicable despite. The model integrates continuous improvement and refinement of solutions based on new data or pretests. The stage-gate approach structures the process, ensuring that each phase has clear checkpoints/milestones before moving forward. This methodology balances the need for systematic progress with the adaptability required in complex socio-technical systems.

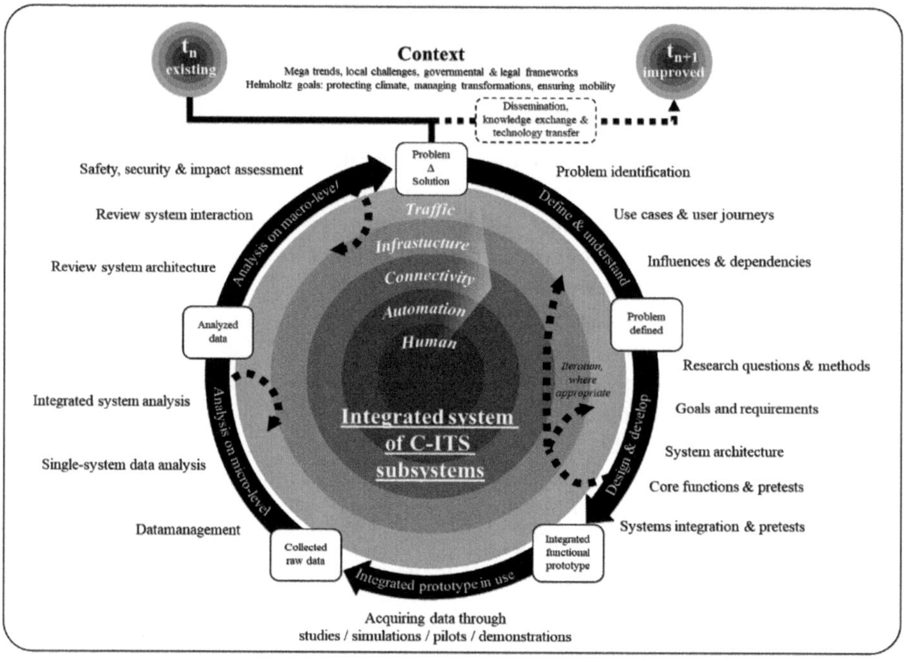

Fig. 5. The new unified C-ITS process model with the phases of Define & understand and Design & develop (preparing the prototype), Integrated prototype in use, and Analysis of the prototype on micro- and macro-level to improve an integrated system (t_n to t_{n+1}).

The *Preparing phase* covers of two critical steps, following the general principles of user-centered design. First, Define & understand focuses on comprehensively understanding the challenges and opportunities within the existing C-ITS system. Conducting a thorough problem definition from all angles (subsystem perspectives) with appropriate methods (e.g., stakeholder analysis, expert interviews, user journey mapping, technical analysis, etc.) leads to gaining insights about the current system's flaws and weaknesses. Achieving a common and contextualized understanding is central to this. Knowledge exchange and collaboration between the subsystems ensures that solutions are informed by the latest technical developments and interdisciplinary expertise. Also, potential (direct and indirect) impact areas and boundaries should be considered early.

The goal is to achieve a joint and robust understanding of a collaboratively defined problem, while taking the system's socio-technical complexities and interdependencies into consideration.

Next, Design & develop starts with formulating very precise research questions, ideally for each subsystem's perspective, leading to system requirements. Furthermore, well established and robust methodological approaches to answer these questions must be identified. Eventually, subsystems work on solutions that aim to answer the research questions, converging into a joint system architecture. Since, depending on the research stage of the iterative cycle, the goal is to develop an integrated functional prototype, that aims to improve the C-ITS system, this stage emphasizes rapid prototyping and iterative refinement, allowing for the testing of proposed solutions (subsystem prototypes) in controlled and individual system contexts. However, system integration is at the core of this stage and the researching teams should define their milestones together. Here, pretests and iterations are key, as they provide opportunities to collaborate, refine, and improve the integrated prototype based on emerging data and practical constraints.

The *Using phase* deploys the integrated prototype depending on project scope. This phase is critical for data collection, as using the integrated system generates valuable raw data based on performance under specific conditions. Using the integrated system reveals whether and to what extend it solves the previously identified problem. Furthermore, this phase will reveal how the interrelated subsystems work together. This step ensures that the developed system can operate as intended, while also identifying any latent issues or optimization opportunities. Depending on the project scope, using the system could vary, for example, from real-world environments as a fully operational system, a demo show case, system-pilot, field study, or simulation. The collected data forms the basis for subsequent analysis and, eventually, further system improvements.

The following *Analyzing phase* divides two complementary levels of evaluation: micro-level and macro-level analyses. Both steps require careful and sustainable data management of the acquired raw data.

The analysis on micro-level involves a detailed examination of the performance of individual subsystems. Goal is to answer the research questions posed to gain scientific knowledge, e.g., regarding performance, validity, but also strengths and weaknesses of the methodological approach. Insights from this analysis are used to improve subsystems and knowledge bases. Furthermore, the analysis of the whole integrated C-ITS system takes a broader view to identify patterns or synergies that may not be apparent when examining subsystems independently. This step evaluates effectiveness and efficiency of the integrated prototype, ensuring that subsystem-level improvements align with the overarching goals and improve the status quo.

Eventually, the macro-level analysis plays a critical role in evaluating the overarching performance of the integrated prototype (the improved target system), its architecture, as well as the subsystem interaction and their interdependencies. Besides validating the internal consistency of the prototype, this step ensures that insights from micro-level analysis are in accordance within the broader C-ITS ecosystem for real-world applicability (external validity), since the overarching goal is the positive impact on mobility and transport systems.

Therefore, the systemic effects of the new prototypical system are very important to evaluate. In that regard, areas for evaluation can span a wide range, such as improving road safety, optimizing transport efficiency, enhancing mobility (also in terms of accessibility), increasing reliability, reducing energy use, reducing congestion and environmental impacts, and even supporting economic development. Also, other impact areas, such as standardization, industry, other user groups, transport authorities, or national and international institutions, as well as governments, could be targeted when assessing the integrated prototype on a macro-level. Hence, the macro-level analysis is essential for identifying emergent behaviors and broader consequences and eventually should provide a decision base for deploying the integrated prototype.

Validating that the developed system cohesively aligns with contextual aspects like societal, ethical and legal, safety/security objectives is crucial, too. The insights gained at this level, altogether with the findings from the micro-level, lead to a comparison of the newly developed and integrated C-ITS system against the status quo.

Consequently, two paths grow from this milestone. One is the next project iteration, where detected insufficiencies guide the next R&D project to improve the integrated system even further. The second is the emphasis on knowledge exchange, dissemination, and technology transfer after a full circle of R&D and knowledge creation. This step is crucial for involved scientific communities, research institutions, but also industry and policy. This post-project activity should facilitate the sharing of insights and innovations, as scientific publications, project reviews, patents, contributions to international standards, open source distributions or eventual spin-offs and licensing of technology.

In summary, our newly proposed unified and user-centered process model for integrated and highly complex C-ITS project serves as a structured framework for the continuous evolution of C-ITS systems. It's systematical integration of iterative refinement and multi-level analysis addresses both the socio-technical and functional dimensions of C-ITS systems. In doing, so the model harnesses the R&D output effectively. This unifying approach enables the transition from an existing HPCS to an improved future state, ensuring sustainable and data-driven progress in C-ITS and beyond.

2.3 Evaluating the New Unified and User-Centered C-ITS Process Model

In order to receive feedback about the unified and user-centered C-ITS process model regarding potentials and limitations in practice, the model was evaluated qualitatively by experts in the field of mobility and transportation research. Therefore, an online focus group session was conducted to discuss strength, weaknesses, as well as potential improvements.

2.3.1 Participants and Procedure

Assessing the proposed new unified and user-centered process model for C-ITS projects, we conducted an expert online workshop to gain qualitative feedback. 4 experts (all male, age $M = 43.25$; $SD = 1.23$; and different from the authors) from the domain of cooperative autonomous driving systems, such as perception and connectivity, traffic infrastructure and automation, attended the session and actively participated in the initial feedback rounds. All participants had an affiliation to the German Aerospace Center

(DLR) and conducted their respective work within the context of transportation systems and automated vehicles. Hence, all participants were familiar with existing process models in R&D and therefore experienced select aspects of the new unified model, depending on their research field and background. Vocational experiences varied (M = 17.75, SD = 0.43) among participants and all where active in large-scale international (EU) and national (German) R&D projects such as GAIA-X, Hi-Drive, InterACTIVE ADAS&ME, ViVre and KIS'M.

The workshop was conducted in an online format with a browser-based collaboration tool and video conferencing software for a duration of three hours. The workshop structure showed four parts, which enabled participants to emphasize the importance for C-ITS projects, familiarize with established process models (V-Model, DevOps Cycle, Spiral Model, User-Centered Design), key challenges in existing processes for R&D in C-ITS from a practitioner's perspective. Eventually, the aim, the requirements, the challenges, and the resulting unified C-ITS process model were presented.

Following this, we initiated a structured work phase, during which we employed a feedback grid to elicit feedback from the participants. For each topic area, a silent brainstorming session followed (Heslin 2009). During the silent brainstorming session, participants recorded their thoughts and comments on individually color-coded virtual sticky notes, which enabled the tracing of a single expert's line of thought across the feedback grid before evoking the open discussion.

3 Results

Following the workshop, the collected data and qualitative feedback by the participants was reviewed and processed in the form of thematic clustering. Accordingly, the initial user feedback was structured and is presented in the upcoming section.

3.1 Potential Upsides

Overall, the participants perceived the unified and user-centered C-ITS process model as a very beneficial tool for potentially enhancing and streamlining R&D activities in complex and multi-facetted C-ITS projects. Here, starting from the user perspective was acknowledged to be very helpful in establishing a common problem statement. The feedback also shows that the circular arrangement of activities provides value to researchers. The highest potential of the process model was considered to be its capability to facilitate cooperation across the teams working on the subsystems within a project, due to joint stage-gates represented in the model. By potentially increasing transparency of the overall working process, participants expect positive effects on the understanding of the current workflows and milestones within the project plan and among involved groups, which was mentioned be tough to handle in current R&D projects. Furthermore, participants indicated that the process model offers a transparent and practical framework to assist developers in navigating and planning their activities along C-ITS projects while maintaining a high quality and punctuality of outputs. This enhanced transparency was perceived to support road mapping and resource allocation of the overall project, enabling developers to strategically map their activities within the superordinate project goals.

This was also seen as a potential advantage when acquiring and initiating new projects, since the model supports the definition of milestones within projects by structuring activities. This aspect could potentially improve the efficiency in grant writing, too.

3.2 Potential Challenges and Suggested Improvements

The early-stage feedback of the new C-ITS process model also revealed challenges like the operationalization of the process, which was not yet defined in this work. To ensure the successful implementation of the C-ITS process model, it seemed essential to establish a shared understanding among stakeholders regarding the interpretation and practical use of the process and each of its steps (iterative stages and gates). The need for a framework that is specific enough to provide clear guidance and applicability, yet flexible enough to accommodate the diverse requirements of various C-ITS projects, scopes, and routines transpired. Furthermore, a need to effectively integrate some sort of "Lessons Learned" or retrospective stage was mentioned in the discussion to enhance adaptability and foster sustainable outcome.

Another challenge identified was the compatibility of diverse tasks in the daily business, e.g., multi project operation, into the process model. For example, in order to successfully meet various requirements of simultaneously conducted projects, it appeared necessary to identify connections across different R&D projects and deliberately define interactions with other process models in use. Highlighting these linkages would help align different task and project deadlines more effectively, enabling smoother and more cohesive integration into prevalent workflows.

Further suggested improvements can be summarized as follows: Participants highlighted the need for additional, more detailed representations of the actual workflow and a detailed interaction of the subsystems. For example, providing a sort of toolbox to enhance the context-specific knowledge required to ensure the applicability of the process model. This proposition could support the mitigation of the previously mentioned challenges of establishing a shared understanding of the process in detail and also improve the establishment of routines into daily R&D activities. A representational adjustment mention was explicitly linking the iteration of the preparation phase (right side of the diagram) with the analysis phase (left side) so that the results of the analysis could directly influence the steps of the next preparation specifically.

4 Discussion

In order to address the high aspirations for future C-ITS research projects and to manage the ever-growing levels of complexity thereof, which is not covered by existing process models, this work aims to come up with a new approach to structure highly integrated R&D projects in C-ITS. This new unified and user-centered process model should improve collaboration and improve effectiveness of R&D output. The here presented process model aims to comply with existing techniques and frameworks to find high adoption rates in the field. Therefore, this work successfully presented both development and evaluation of the new model.

Based on existing and widely acknowledged process models, this work takes a focused approach to combine iterative processes (UCD, Spiral Model, DevOps), verification and validation and system-of-systems approaches (both V-Model), reactivity (UCD, DevOps) and traceability of activities (Spiral Model) and to tailor these for a new process model that aims at managing complexity, and therefore promoting collaboration and research efficiency in the field of C-ITS. Essentially, the new unified user-centered C-ITS process model describes three iterative stages (preparing, using, analyzing) that set the joint framework for R&D activities across all HCPS's subsystems (human-automation-connectivity-infrastructure-traffic) and also give a strategic perspective to the R&D output.

The feedback provided by the workshop participants was very helpful in putting the developed C-ITS process model into perspective of experienced researchers. The insights gained suggest a potentially positive impact for highly complex and deeply integrated R&D projects in C-ITS. This finding is based on the fact that especially the consideration of various subsystems and the iterative approaches could effectively assist the researchers in their activities. Furthermore, the experts valued the opportunity to work on diverse topics and the individual subsystems within a unified stage-gate structure, fostering high levels of cooperation and transparency. In that regard, this work successfully addresses a need in the C-ITS research community. Another key finding was that this new C-ITS process model actively enhances the transition from one project to another from the expert's point of view. By iteratively connecting preparation, using, analysis, and dissemination of a joint R&D context, the activities of each subsystem can be streamlined for time effective, sustainable, and high-quality research results. The proposed structure has the potential to affect dissemination and grant writing positively, too. Consequently, traceability of continuous activities and outcomes within target C-ITS system could be improved, which could also affect strategic planning and resource allocation in future project planning.

However, the conducted expert workshop also uncovered aspects to improve the new C-ITS process model. On the one hand, the iterative approaches and segmentation were perceived as very helpful but a more detailed connection between analysis and iterative preparation, which is seen in some V-Model representations (VDI/VDE, 2021), could further increase the continuous aspect of R&D activities. Another outcome of the expert workshop was that operationalizing each step, e.g. by providing a sort of toolbox, could help to establish the process model among practitioners.

However, this work aimed at streamlining activities across the subsystems, and not disrupting their specific, and well-established methods. This aspect is extremely important to mention, since each research field already has a base of relevant, rigorous research methods. Instead, the unified C-ITS process model aims at organizing these highly complex R&D activities by establishing a mutually helpful base for sustainable collaboration and iterative system integration. In a nutshell, despite the development of the new process model for C-ITS projects being conducted conceptually, the overall feedback was very promising. Hence, the proposed framework could potentially improve C-ITS projects. Although only select aspects, like iterative stages and the integrated system-of-systems approach, were realized in projects, and the new C-ITS process was not tested in its entirety yet, experts already recognized its potential value. Building on these findings, future work should increase the number of experts and let them work with the model first hand to gain more insights, uncover potential areas of conflict or practical inefficiencies, and further update the unified process model. Consequently, steps should be defined more accurately in the near future to comply with the daily work of R&D personnel to make the process model fully applicable (e.g., by generating some sort of practical guidelines or handbooks). Eventually, the proposed new unified user-centered process model for C-ITS projects must stand the test of practicability in actual highly integrated R&D projects over time.

5 Conclusion

The proposed new unified user-centered C-ITS process model represents a significant improvement for transparent, collaborative, and iterative R&D in highly complex research areas such as CCAM. It combines the strengths of common process models, especially in regards to iterations, systems integration, and trajectories of knowledge creation, and blends them into a useful and structured framework. Through the linking of interdisciplinary problem definition and iterative development of an integrated and functional prototype, which is followed by focused data acquisition and analysis/evaluation, this new framework promotes interdisciplinarity and provides a basis for the efficient use of resources. In applying this new framework, C-ITS projects with highly complex and deeply integrated and interdependent subsystems can achieve high-quality research output with traceable results. Eventually, the actual impact of the process model needs to be investigated in the future, since this work only presents a very first conceptual proposal. When applying the model in actual multi-system projects, the impact and the desired benefits could materialize. However, as a first step, this new unified user-centered C-ITS process model could already find utility in less complex projects first, before deploying at large-scale. Concluding, by emphasizing the collaborative integration of all involved subsystems, i.e., from human to traffic, and by harmonizing the iterative stage-gate process for the involved researchers from different perspectives, the unified and user-centered C-ITS process model will potentially foster resource efficiency in large-scale R&D projects in the research field and improve knowledge creation, exchange and transfer. Ultimately, by enhancing the methodology for conducting C-ITS/CCAM projects, current and future challenges in the field can be addressed more effectively. This improved approach contributes to the necessary transformation of the transportation system, focusing on high-level automation, broad connectivity, and overall resource efficiency.

References

Abrahamsson, P., Jedlitschka, A., Duc, A. N., Felderer, M., Amasaki, S., Mikkonen, T. (eds.) Lecture Notes in Computer Science: Vol. 10027. Product-Focused Software Process Improvement: 17th International Conference, PROFES 2016, Trondheim, Norway, November 22–24, 2016: Proceedings. Springer, Heidelberg (2016). https://doi.org/10.1007/978-3-319-49094-6

Amelio, V.D.: Design Interference Detection for Multi-Disciplinary Product Development (2010)

Atlassian: What is DevOps? (2025). https://www.atlassian.com/devops. Accessed 28 Aug 2025

Balalaie, A., Heydarnoori, A., Jamshidi, P.: Microservices architecture enables devops: migration to a cloud-native architecture. IEEE Softw. **33**(3), 42–52 (2016)

Balalaie, A., Heydarnoori, A., Jamshidi, P.: Migrating to cloud-native architectures using microservices: an experience report. In Advances in Service-Oriented and Cloud Computing: Workshops of ESOCC 2015, Taormina, Italy, September 15–17, 2015, Revised Selected Papers 4. Symposium conducted at the meeting of Springer (2016). https://doi.org/10.1007/978-3-319-33313-7_15

Baraldi, P., Di Maio, F., Zio, E. (eds.): Proceedings of the 30th European Safety and Reliability Conference and 15th Probabilistic Safety Assessment and Management Conference. Research Publishing Services (2020)

Barolli, L. (ed.): Lecture Notes on Data Engineering and Communications Technologies. Complex, Intelligent and Software Intensive Systems. Springer, Heidelberg (2024). https://doi.org/10.1007/978-3-031-70011-8

Bass, L., Weber, I., Liming, Z.: Devops: A Software Architect's Perspective: A software architect's perspective. The SEI series in software engineering. Addison-Wesley (2015). https://learning.oreilly.com/library/view/-/9780134049885/?ar

Boehm, B.: A spiral model of software development and enhancement. Computer **21**(5), 61–72 (1988). https://doi.org/10.1109/2.59

Boehm, B., Hansen, W.: The spiral model as a tool for evolutionary acquisition. CrossTalk **14**(5), 4–11 (2001)

Brown, T.: Design Thinking. Harv. Bus. Rev. **86**(84–92), 141 (2008)

Brown, T., Kātz, B.: Change by Design: How Design Thinking Transforms Organizations and Inspires Innovation (1st edn.) Harper Business (2009)

Cheng, L.-T., et al. (eds.): Proceedings of the 2008 international workshop on Cooperative and human aspects of software engineering. ACM (2008)

D'Amelio, V.: Design Interference Detection for Multi-Disciplinary Product Development (Doctoral dissertation, Technische Universiteit Delft) (2010)

Debois, P.: Agile Infrastructure and Operations: How Infra-gile are You? In Agile 2008 Conference (pp. 202–207). IEEE (2008). https://doi.org/10.1109/Agile.2008.42

Deuter, A.: Slicing the V-Model – Reduced Effort, Higher Flexibility. In 2013 IEEE 8th International Conference on Global Software Engineering (pp. 1–10). IEEE (2013). https://doi.org/10.1109/ICGSE.2013.10

Elnur, K., Hashimov, E.: V-Model for air defense systems. Model. Control Inf. Technol. (6), 46–49 (2023). https://doi.org/10.31713/MCIT.2023.011

Erich, F., Amrit, C., Daneva, M.: A qualitative study of DevOps usage in practice. J. Softw.: Evol. Process **29**(6), e1885 (2017)

European Comission: SRIA: strategic research and innovation agenda (2023)

Fink, K., Ploder, C. (eds.): Wirtschaftsinformatik als Schlüssel zum Unternehmenserfolg. DUV (2006). https://doi.org/10.1007/978-3-8350-9122-1

FOT-Net, CARTRE: FESTA Handbook: Version 7 (2018)

George, C.A.: What is user-centred design? In User-Centred Library Websites, pp. 1–23. Elsevier (2008). https://doi.org/10.1016/B978-1-84334-359-2.50001-9

George B.L.: Zhivete Model - A Hybrid of V Model and Agile Scrum for Product Development. INTERANTIONAL J. Sci. Res. Eng. Manage. **07**(11), 1–11 (2023). https://doi.org/10.55041/ijsrem27236

Graessler, I., Hentze, J.: The new V-Model of VDI 2206 and its validation. At - Automatisierungstechnik **68**(5), 312–324 (2020). https://doi.org/10.1515/auto-2020-0015

Graessler, I.: A new V-Model for interdisciplinary product engineering. Universitätsbibliothek Ilmenau (2017)

Heslin, P.A.: Better than brainstorming? Potential contextual boundary conditions to brainwriting for idea generation in organizations. J. Occup. Organ. Psychol. **82**(1), 129–145 (2009). https://doi.org/10.1348/096317908X285642

Humble, J., Farley, D.: Continuous delivery: reliable software releases through build, test, and deployment automation. Pearson Education (2010)

Iden, J., Tessem, B., Paivarinta, T.: Problems in the Interplay of IS development and IT operations: an alignment analysis. In: Pooley, R., Coady, J., Schneider, C., Linger, H., Barry, C., Lang, M. (eds.) Information Systems Development. Springer, New York, pp. 23–33 (2013). https://doi.org/10.1007/978-1-4614-4951-5_3

International Organization for Standardization. (2019). Ergonomics of human-system interaction – Part 210: Human-centred design for interactive systems (ISO 9241–210:2019). ISO

Jeong, J., & Heo, G. (2020). Cyber Security Evaluation for Nuclear I&C Systems Following Software Life Cycle. In: Baraldi, P., Di Maio, F., Zio, E. (eds.) Proceedings of the 30th European Safety and Reliability Conference and 15th Probabilistic Safety Assessment and Management Conference, pp. 2044–2048. Research Publishing Services. https://doi.org/10.3850/978-981-14-8593-0_4843-cd

Lake, J.G.: 4 V & V in plain English. INCOSE Int. Symp. **9**(1), 1134–1140 (1999). https://doi.org/10.1002/j.2334-5837.1999.tb00284.x

Leite, L., Rocha, C., Kon, F., Milojicic, D., Meirelles, P.: A survey of DevOps concepts and challenges. ACM Comput. Surv. **52**(6), 1–35 (2019). https://doi.org/10.1145/3359981

Li, J., Zeng, X., Stevels, A.: Ecodesign in consumer electronics: past, present, and future. Crit. Rev. Environ. Sci. Technol. **45**(8), 840–860 (2015). https://doi.org/10.1080/10643389.2014.900245

Lim, C.H., Chin, J.F.: V-model with fuzzy quality function deployments for mobile application development. J. Softw.: Evol. Process **35**(1), Article e2518 (2023). https://doi.org/10.1002/smr.2518

Lu, G., Lu, S., & Li, X.: Control algorithm development for electric vehicle residual driving distance based on V Model. In: Proceedings of the 2019 International Conference on Modeling, Simulation, Optimization and Numerical Techniques (SMONT 2019). Atlantis Press (2019). https://doi.org/10.2991/smont-19.2019.40

Lwakatare, L.E., Karvonen, T., Sauvola, T., Kuvaja, P., Olsson, H.H., Bosch, J., Oivo, M.: Towards DevOps in the embedded systems domain: Why is it so hard?. In 2016 49th hawaii international conference on system sciences (HICSS), pp. 5437–5446. IEEE (2016)

Mathur, S., Malik, S.: Advancements in the V-Model. Int. J. Comput. Appl. **1**(12), 29–34 (2010)

Mujumdar, A., Masiwal, G., Chawan, P.M.: Analysis of various software process models. Int. J. Eng. Res. Appl. (IJERA) **2**(3), 2015–2021 (2012)

Munassar, N., Govardhan, A.: A comparison between five models of software engineering. Int. J. Comput. Sci. Issues (IJCSI) **7**(5), 94 (2010)

Neely, S., Stolt, S.: Continuous Delivery? Easy! Just Change Everything (Well, Maybe It Is Not That Easy). In: 2013 Agile Conference, pp. 121–128. IEEE (2013).. https://doi.org/10.1109/AGILE.2013.17

Nilsson, A., Wilson, T.: Reflections on Barry W. Boehm's "A spiral model of software development and enhancement". Int. J. Manag. Projects Bus. **5**(4), 737–756 (2012). https://doi.org/10.1108/17538371211269031

Norman, D., Draper, S.W.: User Centered System Design: New Perspectives on Human-computer Interaction. Taylor & Francis (1986). https://books.google.de/books?id=Qz5jQgAACAAJ

Norman, D.: The design of everyday things (Revised and expanded ed.). Basic Books (2013). http://swb.eblib.com/patron/FullRecord.aspx?p=1167019

Nuseibeh, B.: Weaving together requirements and architectures. Computer **34**(3), 115–119 (2001). https://doi.org/10.1109/2.910904

Pomberger, G.: Boehm's spiral model revisited. In: Fink, K., Ploder, C. (eds.) Wirtschaftsinformatik als Schlüssel zum Unternehmenserfolg, pp. 89–98. DUV (2006). https://doi.org/10.1007/978-3-8350-9122-1_5

Pooley, R., Coady, J., Schneider, C., Linger, H., Barry, C., Lang, M. (eds.): Information Systems Development. Springer, New York (2013). https://doi.org/10.1007/978-1-4614-4951-5

Poppendieck, M., Poppendieck, T.: Lean software development: An agile toolkit: An agile toolkit. Addison-Wesley (2003)

Pressman, R.: Software Engineering: A Practitioner's Approach (2014)

Ranjbar Motlagh, R., Ameri Sianaki, O., Shee, H.: A survey on cooperative intelligent transportation systems (C-ITS): Opportunities and Challenges. In: Barolli, L. (ed.) Lecture Notes on Data Engineering and Communications Technologies. Complex, Intelligent and Software Intensive Systems, vol. 87, pp. 253–260. Springer, Switzerland (2024). https://doi.org/10.1007/978-3-031-70011-8_23

Regulwar, G., Gulhane, V., Jawandhiya, P.: A security engineering capability maturity model. In: 2010 International Conference on Educational and Information Technology, vol. 1, pp. V1–306. IEEE (2010)

Rembetsy, M., McDonnell, P.: Continuously Deploying Culture: Scaling Culture at Etsy. Velocity Europe (2012). https://de.slideshare.net/slideshow/continuously-deploying-culture-scaling-culture-at-etsy-14588485/14588485

Riungu-Kalliosaari, L., Mäkinen, S., Lwakatare, L. E., Tiihonen, J., Männistö, T.: Devops adoption benefits and challenges in practice: a case study. In: Abrahamsson, P., Jedlitschka, A., Duc, A.N., Felderer, M., Amasaki, S., Mikkonen, T. (eds.) Lecture Notes in Computer Science: Vol. 10027. Product-focused software process improvement: 17th International Conference, PROFES 2016, Trondheim, Norway, 22–24 November 2016, Proceedings, vol. 10027, pp. 590–597. Springer, Heidelberg (2016). https://doi.org/10.1007/978-3-319-49094-6_44

Rook, P.: Controlling software projects. Softw. Eng. J. **1**(1), 7 (1986). https://doi.org/10.1049/sej.1986.0003

Royce, W. W.: Managing the development of large software systems: concepts and techniques. In Proceedings of the 9th international conference on Software Engineering (1987)

Ruparelia, N.B.: Software development lifecycle models. ACM SIGSOFT Softw. Eng. Notes **35**(3), 8–13 (2010). https://doi.org/10.1145/1764810.1764814

Saunders, M., Lewis, P., Thornhill, A.: Research Methods for Business Students, 8th edn. Pearson (2019). https://elibrary.pearson.de/book/99.150005/9781292208794

Schindler, J., Hesse, T.: Verbesserter/erweiterter Entwicklungsprozess und Toolkette inkl. Entwicklung von Tools zur Verifikation von Anforderungen zusammen mit AP 4200: Meilensteinbericht (2016)

Sharp, H., Rogers, Y., Preece, J.: Interaction Design: Beyond Human-Computer Interaction, 5th edn. Wiley (2019). https://learning.oreilly.com/library/view/-/9781119547259/?ar

Sommerville, I.: Software testing. Softw. Eng. **8**, 537–565 (2010)

Technical Commitee 4.10, V. D. (2022). Development of mechatronic and cyber-physical systems. https://doi.org/10.31224/2452

Tessem, B., Iden, J.: Cooperation between developers and operations in software engineering projects. In: Cheng, L.-T., et al. (eds.) Proceedings of the 2008 International Workshop on Cooperative and Human Aspects of Software Engineering, pp. 105–108. ACM (2008). https://doi.org/10.1145/1370114.1370141

VDI/VDE: Development of mechatronic and cyber-physical systems (2206:2021–11). DIN Media (2021). https://www.dinmedia.de/de/technische-regel/vdi-vde-2206/342674320

Viravan, C.: Lessons learned from applying the spiral model in the software requirements analysis phase. In: Proceedings of the Third IEEE International Symposium on Requirements Engineering, p. 40. IEEE Computer Society Press (1997). https://doi.org/10.1109/ISRE.1997.566838

Part IV: Vehicle Technology Development and Testing

Defining Energy/Emissions Measurement and Testing in the Connected and Automated Vehicle (CAV) Era

Michael Duoba(✉) and Kevin Stutenberg

Argonne National Laboratory, 9700 South Cass, Lemont, IL 60439, USA
{mduoba,kstutenberg}@anl.gov

Abstract. The Automated Road Transportation Symposium (ARTS23) hosted a breakout session titled "Defining Energy/Emissions Measurement and Testing in the CAV Era." This session focused on testing methods for new and emerging connected and automated vehicle (CAV) technology that measure their efficiency and emissions performance. Key stakeholders include consumers interested in fuel economy labels, regulators concerned with CO2 emissions and corporate average fuel economy requirements, and researchers seeking feedback for technology development. The workshop began with presentations from researchers and testing equipment suppliers, discussing existing tools and proposing new potential test procedures. Subsequently, the participants were split into four groups, each tasked with developing a framework for CAV testing. The resulting notional frameworks are detailed in this chapter.

Keywords: Connected and automated vehicles · dynamometer test procedures · automated vehicle energy efficiency · lab testing automated vehicles · road testing automated vehicles · simulation · vehicle-in-the-loop

1 Introduction

Automated vehicles are transitioning from experimental pilot programs to significant vehicle deployment programs. In this transition, OEMs have encountered substantial challenges in the critical energy/emissions testing and certification step in the vehicle development process. Automated vehicles operate under computer-controlled speed regulation and therefore differ significantly from the traditional manual control of human-driven vehicles. Conventional testing standards employ pre-defined human driving patterns (within a driving scenario devised long ago) with specific speed versus time schedules run on a dynamometer to measure the energy consumption in a very controlled manner.

Argonne National Laboratory's work was supported by the U. S. Department of Energy, Office of Energy Efficiency and Renewable Energy, Office of Vehicle Technology under contract DE-AC02-06CH11357. The authors wish to express their gratitude for Danielle Chou's (DOT FHWA Office of Safety and Operations R&D) ongoing support and contributions. Her ideas, guidance, and support have been instrumental in shaping the direction, development, and execution of the workshop.

G. Meyer and S. Beiker (Eds.): ARTSymposium 2024, LNMOB, pp. 127–137, 2026.
https://doi.org/10.1007/978-3-032-12374-9_8

In addition to questions about appropriate deviations to certification testing [1], the automotive research and development community working on CAV-related programs do not have standard ways to evaluate their achievements in a way that can be compared to other R&D efforts. Routes and conditions are developed ad hoc, and often based on available data and specific research objectives. For instance, Michigan Tech researchers performed experiments using a route around campus called the Michigan Tech Drive Cycle (MTUDC) [2], and the presenter from Southwest Research Institute (SwRI) developed cycles internally based on different scenarios and datasets.

The idea for the workshop grew out of discussions between staff at Argonne National Laboratory and the U.S. Department of Transportation's Federal Highway Administration, Office of Safety and Operations Research and Development. While much of the ongoing work in CAV testing has centered on safety and reliability, far less attention has been given to evaluating the potential benefits of CAVs for energy efficiency and decarbonization. The workshop was therefore designed to bring together experts from academia, industry, and national laboratories to share perspectives and identify practical approaches. The ultimate goal was to translate these insights into concrete frameworks for CAV testing that can be further developed and refined.

2 Summary of Presentations

Before the breakout sessions, a panel of subject matter experts provided brief presentations on the problem from their own perspectives from research institutions, academia and test/measurement suppliers. Though automotive original equipment manufacturers (OEMs) were invited and did participate, none chose to present during the segment. Following the presentations, a panel discussion took place, addressing various aspects of CAV testing and incorporating questions from the audience.

2.1 Problem Introduction and Progress in SAE Standards Committee

Michael Duoba, Research Engineer from Argonne National Laboratory provided an introduction and background to the problem. Duoba is the chair of the recently formed *SAE Automated Vehicle Environmental Testing Task Force* which is beginning to tackle the issues discussed in the workshop.

A summary of this presentation includes the following:

- The adage "*You cannot improve what you don't measure*" is frequently cited in both science and business circles. It highlights the necessity of robust and widely accepted tests for making quantitative CAV efficiency assessments because without such tests, improved efficiency may not be considered in their design.
- The advantages of laboratory-based testing under controlled conditions are well established in today's chassis dynamometer vehicle efficiency paradigm. Unless the challenges prove overwhelming, the aim should be to use dynamometer testing for experimental control and repeatability.
- Designing the many trip scenarios to be used in CAV testing presents numerous challenges. These include determining the necessary range of roadway types, assessing varying levels of traffic, and deciding whether existing models are adequate or if new

ones need to be developed to realistically simulate the behavior of human drivers in traffic flow.

2.2 XIL Solutions for Advanced Validation of Vehicle Systems, Perspective from dSpace

Jace Allen, Technology Evangelist, and former Director of ADS/AD Engineering and Business Development for dSPACE, Inc. presented key test equipment enablers for CAVs. Many of the advanced tools and methods used in the CAV development process also have potential use in testing for efficiency and emissions. Existing hardware and software tools that facilitate "software-in-the-loop," "hardware-in-the-loop," and "vehicle-in-the-loop" may offer the flexibility for CAV testing in a controlled laboratory environment. For example, a lab built at Oak Ridge National Laboratory allows testing fully road-capable CAVs using axle dynamometers and a vehicle-in-the-loop simulation testing infrastructure.

2.3 CAVs Can Be Tested "Now" for Energy/Emissions, Perspective by Horiba

Leo Breton, Technology Development Director, HORIBA Instruments provided a practical concept for testing vehicles with Adaptive Cruise Control (ACC). Rather than employing sophisticated vehicle-to-simulation interfaces, the proposed test lab setup includes moving "dummy" vehicle targets that are suspended from above with an actuation system allowing relative movement around the test vehicle that is mounted to a stationary dynamometer. The targets are radar-reflective and appear like a real vehicle from the perspective of the test vehicle sensors. The system tracks and controls the target locations relative to the test vehicle in real-time. Considering many new production vehicles feature ACC, this method could successfully be deployed for ACC testing, eliminating key challenges associated with developing interfaces connecting the simulated environment to vehicle sensors or compute stack.

2.4 Quantifying the Benefits of Energy-Saving Technologies for CAVs, Perspective by Southwest Research Institute (SwRI)

Mike Brown, Institute Engineer at Southwest Research Institute, presented experience with developing and testing a prototype connected and automated vehicle for ARPA-E's NEXTCAR program [3]. The key takeaways include:

- The NEXTCAR project set a clear target—achieve a 30% reduction in energy consumption using Level 4 automation. What remains undefined, however, are the test methods needed to verify this improvement. This gap underscores the need for standardized CAV test procedures, which are critical not only for certification but also for fair and meaningful R&D comparisons.
- SwRI was able to attribute efficiency gains to specific features, including eco-routing, powertrain optimization, eco-driving, cooperative driving, eco-lane changing, and grade optimization. This illustrates the value of evaluating individual features or scenarios separately, since energy savings may vary depending on driving location, operational domain, or traffic conditions.

- Dynamometer tests used speed traces developed from DOE EEMS program corridor simulations. An Eco-driving scenario tested on the chassis dynamometer showed a 16% reduction in energy consumption.
- Preliminary platoon testing at the American Center for Mobility (ACM) showed a 10–12% energy savings when two vehicles traveled in formation at an 8 to 16-m gap at 55 mph.
- SwRI also conducted track testing using augmented reality to simulate surrounding traffic targets. The CAV reacted to the simulated targets in real time, allowing its control system to operate as though the traffic situation was real.

2.5 Michigan Technological University (MTU) NEXTCAR Development and Testing, Perspective by MTU

Bo Chen, Professor of Mechanical Engineering and Electrical Engineering, Michigan Technological University (MTU) provided information about their university's NEXTCAR project. The key takeaways from their project include:

- Using a Chevrolet Volt (2nd Generation) as the platform for their CAV features development, energy consumption was reduced by 20% testing on local roadways. Again, the lack of standard conditions or roadway characteristics for testing was encountered.
- The project follows a path of increased optimization spanning short term, mid-term, and long-term features.

  Short-term features: optimized torque split power train operation, optimized powertrain mode selection, and optimal velocity profiling
  Mid-term features: eco-speed and speed harmonization with other vehicles
  Long-term features: reducing total energy of a route by optimizing blended operation for an optimized eco-routing algorithm.

- Ultimately, by employing modeling + optimization + connectivity, they aim to reduce energy consumption by 30%.
- MTU has a number of CAV development vehicles, including PHEVs (Volt and Chrysler Pacifica), EV (Chevrolet Bolt), and HEV (Ram 1500 mild hybrid) which were built with partners from New Eagle, AutonomousStuff and Pilot Systems.

2.6 Automated and Connected Mobility Testing Perspectives by AVL

Joshua Orlando, Project Engineer II in Connected and Automated Mobility at AVL, presented a range of existing testing systems and tools useful for CAV testing aimed at energy efficiency. In addition to a proving ground with a smart city module for ADS testing and the indoor Mobility and Sensor Center for sensor benchmarking, AVL's DRIVINGCUBE™ provides an integrated testing setup for automated vehicles. This system, a complete Vehicle-in-the-Loop solution, includes an array of hardware that can stimulate and/or make data injections for on-board sensors such as: radar, cameras, ultrasonic sensors, accelerometers, GNSS, and Lidar. A unique element in the setup is a steering force emulator, enabling testing on traditional chassis dynamometers while accommodating steering actuation with proper auxiliary loads. This platform could serve as a valuable starting point for testing professionals to explore CAV testing concepts and the potential standardization of methods and hardware.

2.7 Panel Discussion

In the panel discussion, moderators highlighted some critical components needed for the development of a testing methodology. The first of which addresses the critical need to **fully capture all the CAV features** that require testing to determine their impact on economy and emissions. Specific questions posed in this discussion included:

- Should we create a portfolio of individual testing scenarios tailored to each CAV feature or group, or should a single, all-encompassing test be used?
- In testing the energy efficiency of automation features, should connectivity be integrated into the assessment, or treated as a distinct category?

Secondly, the defined CAV features need to be tested with **representative scenarios** that determine a features potential for energy savings. Questions for this discussion included:

- What are the best practices for developing standardized scenarios (or general tests) to accurately reflect real-life operational domains?
- Is it necessary to include large-scale data collection from on-road operations as part of scenario development?
- Should we integrate new scenarios that mimic established drive cycles (like U.S. EPA UDDS, HWFET, etc.)? If so, what would be the method for this integration?
- What strategies can be employed to capture and manage randomness in the test scenarios?

Thirdly, we must consider defining the essential tools and testing processes that constitute the **testing framework**. This involves addressing a final set of questions, which include:

- How should we balance simulation with physical testing to achieve the most effective outcomes?
- What strategies do you recommend for integrating vehicles into these scenarios or simulated environments to ensure accurate and relevant testing?
- Is it necessary to require manufacturers to provide a 'test mode' (similar to current dyno mode) that allows for specific testing functions within these scenarios? How would this impact the testing process?
- What are the potential impacts of Over-the-Air (OTA) updates on the testing framework, and how can these be managed or incorporated into testing protocols?

3 Breakout Sessions

The participants were organized into four distinct breakout groups, each tasked with completing two specific assignments within a one-hour timeframe. The groups were able to choose how much time they spent on each assignment.

The initial task involved responding to survey questions designed to prompt the groups to consider a variety of current open issues critical to CAV energy/emissions testing (details in Sect. 3.1).

The second task involved designing a test procedure process flow, allowing each group freedom to develop their own testing framework. The choice of automation level was also determined by each group. Group 1 chose to focus on vehicles with ADAS, targeting lower levels of automation. The other three groups focused on testing connected vehicles with Level 3 + automation. Each group illustrated their test process with a functional flow diagram on a "Jamboard," a cloud-based collaborative tool from Google (details in Sects. 3.2, 3.3 and 3.4).

3.1 Survey Results and Initial Discussions

This subsection discusses the key findings from the surveys distributed to the work-groups. An analysis revealed a prominent theme: the significant variation in responses underscores the daunting challenges in developing future CAV test procedures.

Q1: What is the role and relationship between different types of testing: simulation? Dynamometer (dyno)? track? On-road? Real-world large-scale driving datasets ("Big Data")? Other?

> The responses confirmed that it may depend on the application, which included "*simulation can't be done in a vacuum, needs to be informed by dyno and track testing. Real-world Big Data can be beneficial but highly dependent on interpretation.*" Other responses suggested to "*start with simulation - then dyno - then track/road testing to validate simulation results and do final certification.*"

Q2: Should the test design consist of one comprehensive test or multiple smaller, specialized tests?

> The survey responses predominantly suggested that a lengthy, singular test is impractical. However, there was agreement that the appropriate test length depends on the testing type. For instance, on-road testing scenarios could be conducted through an extended, unified test, which might prove to be effective.

Q3: What are the outcomes expected from the energy/emissions test? Each group selected a different answer:

- *Difference in energy use compared to the standard United States Environmental Protection Agency (EPA) drive cycles.* Does this answer imply that the scenarios must resemble traditional EPA cycles?
- *Difference in energy use compared to human driver performance in the same scenario.* This answer implies a requirement for replicating the test scenario with a human driver, in addition to the connected and automated test conditions. The question then arises: does this additional test necessitate a real human driver, or can a model sufficiently represent the baseline results? What human driving model is suitable? Can that model be standardized?

- *Total energy use measured in a manner that is comparable across various designs.* A standard test would allow for results that are comparable across different vehicles subjected to the same test. Additionally, it would enable the possibility of repeating the test with the CAV features turned off, to assess the effectiveness of these features.
- *Total energy usage.* Is there a baseline? Is there a need to factor human driver results (addressed in the next question)?

Q4: Is it possible to perform a human baseline model or test using the same trip and/or Operational Design Domain (ODD)? This question is critical. If there were a robust and universally accepted model for human driving behavior, it would simplify resolving outcomes in any of the testing and/or simulation frameworks. The responses varied: one group affirmed it's possible, another suggested it might be feasible but requires further investigation, a third group thought it not possible, and the last group believed it's relevant only to Level 5 automation.

Q5: What data do we need? Where might we find that data? The subsequent discussion focused on data requirements. Consensus was reached on the necessity of gathering speed/time profiles, traffic flow, fleet data, as well as connectivity information from both vehicles and infrastructure. There was a unified view that "*Fleets and manufacturers need to collaborate to record and store/aggregate data. Onboard measurement from CAN or other vehicle networks, third party anonymous collection required, for regulatory likely need non-OEM measurement of energy/speed profiles.*" Other suggestions included "*statistical data for nominal driving conditions*" and "*Data captured by vehicle, provided either via over-the-air data collection (like OnStar) or OBD logged / broadcast data.*"

Q6: What tools are required? The responses included:

- Tools suitable for every testing environment are crucial, necessitating a diverse mix. This includes the use of standards and common definitions as essential tools.
- Software tools, simulation tools, instrumentation and dynamometers.
- Interface tools like radar injection, camera screen or injection, radar emulator, steering force emulator.
- Means for large data collection and processes for securely logging data from a fleet of vehicles.

Q7: Do the tools already exist? The answers were essentially yes. On-board diagnostic (OBD) data collection can be used. But many need to be developed or adjusted to work for energy centric applications because much of the CAV data collection that exists was originally designed for safety and robustness.

Q7: Should testing tools be open-source? The answers were unanimously yes.

Q8: Should the scenario/environment respond in real-time to vehicle behavior? This question relates to how the testing framework is conducted. If there is no feedback from the environment, then the test is run by "playing" a pre-determined test scenario much like the current time-speed driving schedules. However, if the environment (primarily other vehicles in-theater) does respond, then each agent must be modeled with sophisticated agent models that aim to accurately represent real-life behavior. The answers were mixed:

- Yes, they should. To assess vehicle closed-loop response. (2 groups)

- No, open-loop playback of the scenario is sufficient to assess vehicle performance (1 group)
- N/A, Breakout group focused on on-road data collection so the question doesn't apply to their framework (1 group)

Q9: Provide additional comments: Feedback suggested that the survey's length was somewhat long considering the available time, particularly since the questions required substantial discussion and decision-making. An important observation was that responses vary significantly depending on whether the objective was certification testing or research and development testing. An important distinction in CAV testing but not as critical for current chassis dynamometer energy and emissions testing procedures.

3.2 Testing Framework Results: Group 1 (ADAS)

The Group 1 framework is presented in Fig. 1. This framework starts with a data collection effort that forms the foundational components of testing ADAS systems (the circled portion in Fig. 1). A large collection of transportation system data will be required to define representative roadways and traffic scenarios. In addition, there is a need for data to help develop high-fidelity human-driven agents; not only for the surrounding traffic, but also to repeat the trip in the test vehicle with a baseline human driver. The required data can come from vehicles driving in the real-world under normal use ("Big Data" sets) along highly diverse types of roadways.

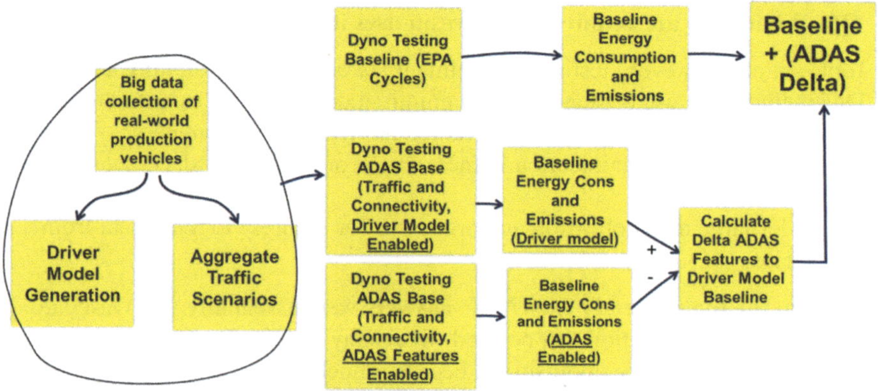

Fig. 1. Breakout Group 1 Framework – focus on CAV efficiency credit applied to existing consumption and emissions testing methods.

Once the testing scenarios and human agent driving models are developed, the aim is to run two parallel CAV testing paths: one with CAV feature enabled, and one without. Testing occurs on a chassis dynamometer equipped with all necessary hooks and communication links to deliver the required environmental stimuli, enabling the vehicle's control system to operate as though it were on the roadway depicted in the testing scenarios.

This approach does not replace standard certification testing which establishes vehicle efficiency and range. The final CAV results provide the percent magnitude of credit (or penalty) applied to the existing vehicle certification results.

3.3 Testing Framework Results: Group 2 (CAV)

The Group 2 framework, illustrated in Fig. 2, prioritizes defining the Operational Design Domain (ODD) constraints and establishing the testing scope for each test vehicle as the preliminary steps. The framework highlights methods for testing a fleet of CAVs to quantify energy consumption. Initially, baseline, conventional chassis dynamometer testing of EPA cycles is required. Results are also simulated to ensure each test vehicle has an equal mix driving situations and scenarios. The prime activity is collecting data from real-world operation. Many operational metrics are collected and aggregated. Once the real-world data is analyzed, the metrics are simplified to provide a final result.

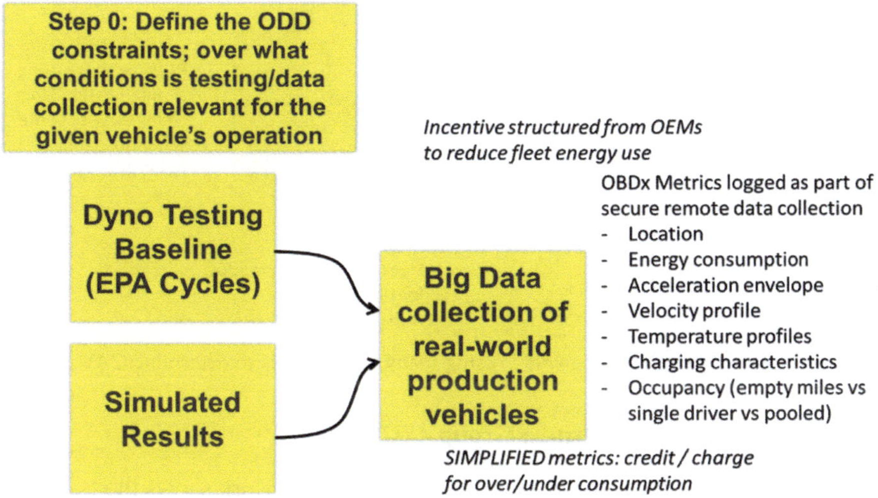

Fig. 2. Breakout Group 2 Framework – focus on fleet performance from on-road testing.

The group noted that if the results are not available to the public (like the consumer fuel economy label displayed on all new cars in the U.S.), then there should be some mechanism structured to provide incentives for the OEM to reduce energy usage on the road. Feedback from on-road data to the simulation effort could also enable the standardization of routes and scenarios, ensuring a level playing field among various autonomous vehicle deployments.

3.4 Testing Framework Results: Group 3 (CAV)

The Group 3 framework is presented in Fig. 3. This framework provides roles for a wide array of testing tools. Laboratory dynamometer testing is used to run common trips/scenarios and satisfy regulatory activities. These tests are run "closed-loop" where the vehicle control system is operating as if it were being driven on a roadway. Track testing is used to characterize and validate behavior and energy usage trends. The group also highlighted ways to run a standard test on a track. The test vehicle is configured on the track to follow a lead vehicle, which drives according to the standard EPA speed trace. This track procedure facilitates a controlled, standard test scenario without linking to sophisticated sensor stimulation hardware. Simulation can be used during development when iterative changes are made, and the energy impacts need to be quantified. Big Data collection can also be utilized to gain insights into specific vehicle fleets for fleet operators and/or municipalities.

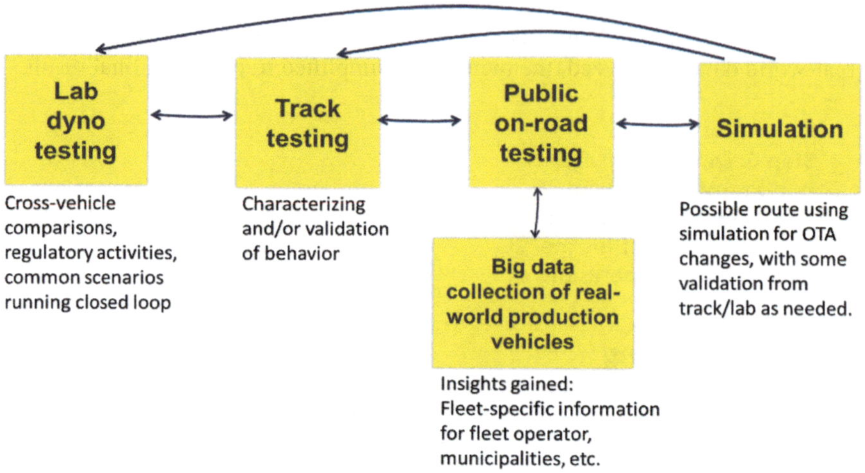

Fig. 3. Breakout Group 3 Framework – using many testing tools to determine CAV efficiency

3.5 Testing Framework Results: Group 4 (CAV)

The Group 4 framework is presented in Fig. 4. This framework shows the testing tools with many possible flow paths for the purposes of helping guide design and development processes. One flow path starts with simulation and moves to laboratory testing, then track and then to public road testing. These testing regimes are focused on the specific scenarios that show the performance of a test vehicle's specific CAV features. In addition, another path of data flows from uncontrolled in-use driving which is processed to find the relevant performance behavior of the CAV. The group pointed out that the process must repeat for each new design or change in design.

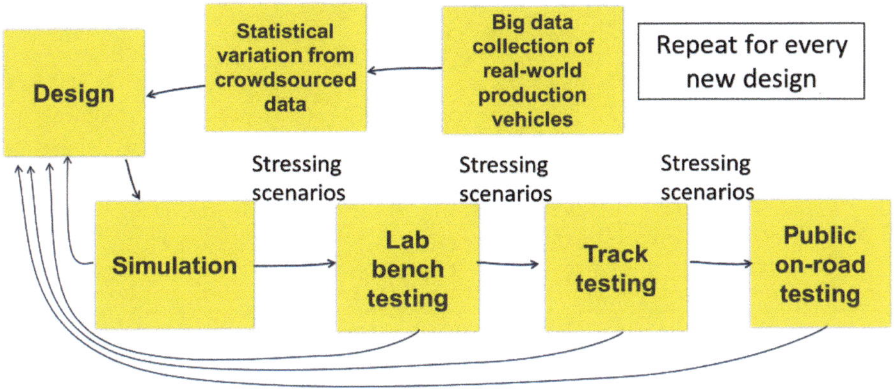

Fig. 4. Breakout Group 4 Framework – focusing on testing for design and development purposes

4 Conclusions and Next Steps

The key takeaway from the "Defining Energy/Emissions Measurement and Testing in the CAV Era" workshop is that no single CAV test method emerged from the discussions that could serve the diverse set of vehicles and testing needs. Multiple technology dependent paths with additional options are required moving forward. It's clear that the procedure development process must be iterative, incorporating simulation, lab, track, and real-world testing. A critical distinction exists between research and development and production/certification testing and this will require innovative ideas and careful development to determine whether both types of testing can be satisfied by a single procedure.

While comprehensive use of many tools can yield very accurate results, the proposed frameworks are significantly more complex and time-consuming than current certification tests. An OEM representative highlighted the impracticality and high costs of some proposed methods. This underscores the need for a collaborative development process to establish tests with more reasonable levels of complexity that still yield characteristic results.

Moving forward, the focus should be on both 1) using available tools to develop methods to test and validate CAV efficiency/emissions performance, but also 2) develop less burdensome approaches that maintain fidelity with the comprehensive approaches to ensure broad acceptance by the testing community.

References

1. United States Environmental Protection Agency. CD-2022-13: (LDT, LDV) Testing of Autonomous Battery Electric Vehicles, August 22, 2022 (2022). https://dis.epa.gov/otaqpub/display_file.jsp?docid=55592&flag=1
2. Oncken, J., Orlando, J., Bhat, P.K., Noordhoek, B. et al.: A connected controls and optimization system for vehicle dynamics and powertrain operation on a light-duty plug-in multi-mode hybrid electric vehicle. SAE Technical Paper 2020-01-0591 (2020). https://doi.org/10.4271/2020-01-0591

Author Index

G. Meyer and S. Beiker (Eds.): ARTSymposium 2024, LNMOB, p. 139, 2026.
https://doi.org/10.1007/978-3-032-12374-9

GPSR Compliance

The European Union's (EU) General Product Safety Regulation (GPSR) is a set of rules that requires consumer products to be safe and our obligations to ensure this.

If you have any concerns about our products, you can contact us on ProductSafety@springernature.com

In case Publisher is established outside the EU, the EU authorized representative is:

Springer Nature Customer Service Center GmbH
Europaplatz 3
69115 Heidelberg, Germany

Batch number: 10755332

Printed by Printforce, the Netherlands